Experiment!

Experiment!

Planning, Implementing and Interpreting

Öivind Andersson

Lund University, Sweden

A John Wiley & Sons, Ltd., Publication

This edition first published 2012
© 2012 John Wiley & Sons, Ltd

Registered office
John Wiley & Sons Ltd, The Atrium, Southern Gate, Chichester, West Sussex, PO19 8SQ, United Kingdom

For details of our global editorial offices, for customer services and for information about how to apply for permission to reuse the copyright material in this book please see our web site at www.wiley.com.

Library of Congress Cataloging-in-Publication Data

Andersson, Öivind, 1970–
  Experiment! : planning, implementing, and interpreting / Öivind Andersson.
    p. cm.
  ISBN 978-0-470-68826-7 (cloth) – ISBN 978-0-470-68825-0 (pbk.)
1. Research.   2. Experimental design   3. Science–Methodology.   I. Title.
  Q180.A1A625 2012
  001.4′34–dc23

                                                              2012008855

A catalogue record for this book is available from the British Library.

Print ISBN: Cloth: 9780470688267 Paper: 9780470688250

Typeset in 10/12pt Times by Aptara Inc., New Delhi, India
Printed and bound in Malaysia by Vivar Printing Sdn Bhd

Instructors can access PowerPoint files of the illustrations presented within this text, for teaching, at:
http://booksupport.wiley.com

1   2012

For Carl-Johan,
an avid experimenter

Contents

Preface   xi

Part One   Understanding the World

1   You, the Discoverer **3**
    1.1   Venturing into the Unknown   4
    1.2   Embarking on a Ph.D.   5
    1.3   The Art of Discovery   5
    1.4   About this Book   7
    1.5   How to Use this Book   8
    Further Reading   10
    References   10

2   What is Science? **11**
    2.1   Characteristics of the Scientific Approach   11
    2.2   The Inductive Method   14
    2.3   The Hypothetico-Deductive Method   16
    2.4   Consequences of Falsification   19
    2.5   The Role of Confirmation   21
    2.6   Perception is Personal   23
    2.7   The Scientific Community   29
    2.8   Summary   30
    Further Reading   31
    References   31

3   Science's Childhood **33**
    3.1   Infancy   33
    3.2   Ionian Dawn   34
    3.3   Divine Mathematics   38
    3.4   Adolescence – Revolution!   41
    3.5   The Children of the Revolution   47
    3.6   Summary   50
    Further Reading   50
    References   51

4   Science Inclined to Experiment **53**
    4.1   Galileo's Important Experiment   54
    4.2   Experiment or Hoax?   56
    4.3   Reconstructing the Experiment   58
    4.4   Getting the Swing of Things   60

4.5   The Message from the Plane 62
4.6   Summary 63
References 64

5   Scientists, Engineers and Other Poets **65**
5.1   Research and Development 65
5.2   Characteristics of Research 68
5.3   Building Theories 70
5.4   The Relationship between Theory and Reality 75
5.5   Creativity 77
5.6   Summary 79
Further Reading 80
References 80

Part Two   Interfering with the World

6   Experiment! **83**
6.1   What is an Experiment? 83
6.2   Questions, Answers and Experiments 85
6.3   A Gallery of Experiments 88
6.4   Reflections on the Exhibition 108
6.5   Summary 110
Further Reading 110
References 112

7   Basic Statistics **113**
7.1   The Role of Statistics in Data Analysis 113
7.2   Populations and Samples 115
7.3   Descriptive Statistics 116
7.4   Probability Distribution 122
7.5   The Central Limit Effect 126
7.6   Normal Probability Plots 129
7.7   Confidence Intervals 132
7.8   The *t*-Distribution 134
7.9   Summary 136
Further Reading 137
References 138

8   Statistics for Experiments **139**
8.1   A Teatime Experiment 139
8.2   The Importance of Randomization 141
8.3   One-Sided and Two-Sided Tests 142
8.4   The *t*-Test for One Sample 143
8.5   The Power of a Test 148
8.6   Comparing Two Samples 150
8.7   Analysis of Variance (ANOVA) 155
8.8   A Measurement System Analysis 159
8.9   Other Useful Hypothesis Tests 163

8.10 Interpreting *p*-Values. 164
8.11 Correlation 165
8.12 Regression Modeling 167
8.13 Summary 171
Further Reading. 172
References 173

9 Experimental Design. **175**
9.1 Statistics and the Scientific Method. 175
9.2 Designs with One Categorical Factor 176
9.3 Several Categorical Factors: the Full Factorial Design 178
9.4 Are Interactions Important? 186
9.5 Factor Screening: Fractional Factorial Designs 187
9.6 Determining the Confounding Pattern. 188
9.7 Design Resolution. 190
9.8 Working with Screening Designs 191
9.9 Continuous Factors: Regression and Response Surface Methods 195
9.10 Summary 207
Further Reading. 208
References 209

10 Phase I: Planning **211**
10.1 The Three Phases of Research 211
10.2 Experiment 1: Visual Orientation in a Beetle 213
10.3 Experiment 2: Lift-Off Length in a Diesel Engine 216
10.4 Finding Out What is Not Known 218
10.5 Determining the Scope. 221
10.6 Tools for Generating Hypotheses. 222
10.7 Thought Experiments 227
10.8 Planning Checklist. 229
10.9 Summary 231
References 233

11 Phase II: Data Collection. **235**
11.1 Generating Understanding from Data. 235
11.2 Measurement Uncertainty 236
11.3 Developing a Measurement System. 238
11.4 Measurement System Analysis. 244
11.5 The Data Collection Plan. 248
11.6 Summary 251
Further Reading. 252
References 252

12 Phase III: Analysis and Synthesis **253**
12.1 Turning Data into Information 253
12.2 Graphical Analysis. 256
12.3 Mathematical Analysis. 259
12.4 Writing a Scientific Paper 260

12.5 Writing a Ph.D. Thesis.   264
12.6 Farewell.   266
12.7 Summary   266
Further Reading.   266
References   267

Appendix **269**
Standard Normal Probabilities   269
Probability Points for the *t*-Distribution.   270

Index **271**

Preface

Today, externally funded research is the rule rather than the exception. The problem statements and hypotheses of a research project have already been formulated by a senior researcher when filing the funding application. When the research student enters the picture a project plan has already been made. All that remains is to produce results in the laboratory. This situation enhances the impression that many students have, that experiments are mainly about taking measurements – a belief arising during traditional, structured laboratory classes that are based on demonstrations rather than real experiments.

Experienced experimenters know that the main efforts in a successful experiment lie in meticulous planning and analysis, and that both of these must be firmly rooted in a scientific approach. Entering the research process at such a late stage, students may be delayed in realizing this crucial point. The art of experimentation is often learned by doing, so an intuitive understanding of the experimental method usually evolves gradually during years of trial and error. The aim of this book is to speed up the journey.

Research education is often practically oriented, focusing on techniques and concepts that are used in specific research areas. In order to become an independent researcher it is important to also acquire more general research competencies. Research students must adopt a scientific mindset, learn how to plan meaningful experiments, and understand the fundamentals of collecting and interpreting data. This book focuses on these general research skills. It is directed to anyone engaged in experiments, and especially Ph.D. and master's students just starting to develop their own experiments.

There are several books that discuss the scientific method from a philosophical standpoint. There are also good books covering research methods from a statistical point of view. Why do we need another book on scientific method? The reason is that scientists need to employ several types of research competency in parallel, and it is unfortunate that these tend to be taught separately. The philosophy of science is frequently introduced using historical examples in courses where more time is spent criticizing ideas than explaining how to apply them. It is too common to teach statistics as a pure study of numbers without explaining how numbers may be turned into meaningful scientific knowledge. The lack of a context relevant to working scientists makes it difficult for research students to incorporate such knowledge into their other research competencies. This book aims to merge these aspects of research into a practically workable package.

The book is organized in two parts. The first part gives a general introduction to the scientific approach while the second describes general methods and tools with a focus on experiments. Towards the end of the book a methodology is presented, which leads the reader through the three phases of an experiment: "Planning", "Data Collection", and "Analysis and Synthesis". The first phase puts the discussion about scientific and experimental method into a practical context, treating how research problems may be identified and how to approach them by conceiving meaningful experiments. The following two phases continue to build on these ideas, but incorporate statistical techniques into the process. In the data collection phase, a measurement system is devised, analyzed and improved to obtain data of sufficient quality. In the last phase, the raw data are turned into meaningful information, analyzed with graphical and mathematical tools, and connected back to the research question.

It is my experience that this methodology helps students connect the somewhat abstract concepts from statistical theory and the philosophy of science to their scientific praxis. I am not an advocate of cookbook recipes for research and the methodology is not to be viewed that way. The idea is rather that readers, by applying these simple tools to their own research problems, will reflect on and learn to understand key elements of the experimental method.

Many people have kindly helped me in the preparation of this book. My friend and colleague Rolf Egnell deserves a very special thank you for taking the time to read and comment on the manuscript in its entirety. He has had the admirable ability to look beyond the depths of his knowledge and read the text with the eyes of a layman. His valuable comments have significantly improved the readability of this text. I am also grateful to Tim Davis for his comments on the statistics material, and to Staffan Ulfstrand for providing feedback on parts where evolutionary biology is discussed. Advice and help of various kinds were willingly given by Malte Andersson, Marie Dacke, Leif Lönnblad, Carl-Erik Magnusson, Clément Chartier, Johan Zetterberg, and Tony Greenfield. I also want to thank Sarah Tilley at Wiley in Chichester, UK, for her support and encouraging remarks along the way. I gratefully acknowledge these people while stressing that I am, of course, solely responsible for any mistakes or other authorial shortcoming still present in the text.

When writing this, others come to mind that have helped me over the years. Many have had a hand in teaching me to see the world around me and to think independently, but space prevents me from mentioning them all. I especially owe thanks to the teachers and staff at the Departments of Physics and Theoretical Physics in Lund, who always presented scientific ideas with a perfect blend of playfulness and intellectual sharpness. I am also grateful to my mother for her unconditional support in all life's endeavors. As a child she instilled in me that I succeeded with everything I tried my hand at, even though I often did not.

Above all, my deepest and most heartfelt gratitude goes to Gunnel, the love of my life, and Carl-Johan, our son, for their sympathy, patience and laughter. You are the twin star around which my world blissfully revolves.

Part One
Understanding the World

1 You, the Discoverer

All truths are easy to understand once they are discovered; the point is to discover them.
—Galileo Galilei

First of all, I would like to congratulate you. The reason that you are reading this book must be that you are learning about scientific method. This should open up possibilities for a variety of interesting professional tasks in the future. Scientific method can be said to be an approach for breaking a complex problem down into its essential components, investigating these components through relevant data and critical thinking, and finally solving the main problem by putting the components back together. Being familiar with scientific method makes it easier to handle complex problems, wherever we encounter them. This is of course useful in various branches of scientific research, where the method was developed, but complex problems occur in many situations. It does not harm policy makers, managers, or technical experts, for example, to be skilled critical thinkers, creative problem solvers, and to be able to draw relevant conclusions from data.

As conveyed by its title, this book is written for experimenters. Experiments are made in a wide variety of research and development tasks, to test ideas and to find out how things work. The book grew out of a need to teach research students general research skills. This means that it is mainly directed towards Ph.D. students, but it is likely also that others working with research and development will find it useful. Having worked with technical development myself, I know that planning, conducting and interpreting experiments are crucial for engineers too. I hope that no one will feel excluded by the way the text addresses research students because most of the contents are very generally applicable. Most readers will probably be found in various parts of the natural, engineering and medical sciences. The reason why I think that the book is relevant across so many disciplines is that the experimental method is such a general approach to finding things out.

Experiments are our most efficient method of discovery. They are based on interfering with something to see how it reacts and find out how it works. The alternative to this is passive observation – standing back, watching and waiting for something interesting to happen. Passive observers obtain large amounts of data, but it may be difficult for them to isolate the information that is relevant to a particular research problem. Experimentation, on the other hand, provides us with data that are tailored to our problem. This means that experimenters have to spend less effort sorting out irrelevant information.

Experiment!: Planning, Implementing and Interpreting, First Edition. Öivind Andersson.
© 2012 John Wiley & Sons, Ltd. Published 2012 by John Wiley & Sons, Ltd.

1.1   Venturing into the Unknown

Research studies are a voyage into the unknown in at least two respects. Firstly, as a Ph.D. student you will often find yourself at the borders of our current knowledge, spying out into the uncharted territories beyond. Secondly, research studies involve learning in ways that are unfamiliar compared to the previous stages of your education. As will be explained in later sections, this is because research studies are focused on acquiring skills in addition to knowledge.

Science is probably the most remarkable enterprise that humankind has ever undertaken. During the couple of hundred thousand years that our species has existed, almost all the knowledge we have about the world we live in has been established during the last couple of hundred years – a mere thousandth of our history. We have been intelligent and curious for eons but only learned to understand Nature once science was invented. This says something of how powerful this idea is. Besides the knowledge it has brought with it, science has also fundamentally changed our living conditions. Scientific progress has found applications in technology and medicine, and these have opened up new avenues of applied research. Science is so integrated in the progress of society that, today, it is all but impossible to imagine an existence without it. This is probably one reason that science is held in such high esteem and that scientists have such high credibility among people.

I would like to clarify that the word science, in this book, refers to the natural sciences in a broad sense. This is not to say that other branches of research are of lesser value, but the methods of science differ between disciplines and the definition has simply been chosen to fit the audience. Many methods of natural science are shared between the natural, engineering and medical sciences. The word "science" itself has Latin roots and originally means knowledge, but it is generally used in a wider sense to include both scientific research and the knowledge that results from it.

With a bit of poetic license you could say that science is a state of mind. Of course, it consists of various facts and methods, but it is also an attitude to the world around us. Scientists act from the presumption that the world is understandable. They try to look beyond the apparent disorder and search for patterns. From these they identify the underlying, general rules that govern the world. This attitude is the basis of scientific discovery.

Do not be fooled into believing that there are no more discoveries to be made. Even relatively simple questions may direct us to fascinating insights about the world. Have you ever wondered why the sky is blue or why the grass is green? The blue sky is now well understood, thanks to the nineteenth century discovery of a phenomenon called Rayleigh scattering. We will return to it in Chapter 10, in connection with a nice experiment about how an African dung beetle finds its way at twilight. The color of the grass, on the other hand, is not so well understood. There is, of course, the obvious reason that grass contains chlorophyll, which absorbs sunlight for the photosynthesis. But it seems odd that chlorophyll should be green because this means that it reflects light in the green wavelength region, where the sun's radiation contains most energy. A black plant, absorbing all visible wavelengths, would extract more energy from the sunlight than a green one. Why do plants reject the sun's energy where it is most abundant? When asking biologists about this, they usually embark on long, interesting discussions about photosynthesis in various organisms before wrapping it up with "So, it's really a good question!" The fact seems to be that we simply do not know.

When you start doing research you will find that there are plenty of unanswered questions in your field as well. Towards the end of this book we will discuss how you may go about finding, approaching and answering them. There are probably more unanswered questions in research today than ever before. As our knowledge grows, the boundaries of knowledge grow too. This means that many discoveries remain to be made, and some of them are waiting for you.

1.2 Embarking on a Ph.D.

To take a systematic approach to discovery you need to acquire new skills as well as new knowledge. Undertaking a Ph.D. will, hopefully, be one of the most developing periods in your life. In all fairness, it will be challenging too. You are likely to work alone a good part of the time. Another challenge is that you will be learning in a completely new way. It may be a shock to discover that high grades from previous studies do not necessarily help you now. The educational system often tends to assess our ability to learn facts and techniques rather than our actual problem solving skills, and research studies require more than having a good head for studying. When you embark on a Ph.D. you go from being a consumer of knowledge to a producer of knowledge; this requires a new set of talents. You do not obtain a Ph.D. for what you know but for what you can do. In other words, research education is largely about acquiring skills and this is why it takes time.

Undertaking a Ph.D. involves a more holistic approach to learning than your previous studies did. Schoolteachers often tend to teach their subjects separately. In high school, for example, it took my fellow students and I a good while to understand that calculus could be used to solve physics problems because it was taught without any hints to applications in other subjects. Little did we know that calculus had been developed by physicists in order to solve physics problems! This tradition of teaching often continues at university. The title of my undergraduate statistics book, for instance, announced that it was written for scientists and engineers. Despite this it did not contain a single example from science or engineering. I often wonder how students are expected to be able to apply subjects that are taught devoid of a relevant context. It is like teaching carpentry by handing out toolboxes and waiting for the students to discover how to use them.

From this point on, things will be different. During your research studies, you will acquire skills by engaging in real research tasks under the supervision of a professional researcher. This is similar to how carpentry is actually taught: if you are to learn how to build a house, you build a house to learn it. Although supervision and textbooks are required, skills can only be learned through hands-on experience. As a research student you will learn about research methods, measurement techniques, and statistical analyses by applying them to real research problems.

1.3 The Art of Discovery

Methods and techniques are all very well but good research involves more than that. As previously stated, science is also an attitude to the world. Professional researchers need to have a scientific mindset. When graduating you should be able to identify and formulate new research questions, develop your own approaches to answering them, and independently conduct research that meets the accepted quality standards of your field. In other words, you should master the scientific method to the point that you are able to make your own contribution to science.

To do this, you need a competency that is made up of several components. One of them is *knowledge*. In order to identify interesting research questions, you must have broad knowledge within your field as well as deep knowledge within the limited part where you are conducting your research. Another component is *skills*, meaning that you need to master a number of methods and techniques. Some of these will be quite general while others will be specific to your field, such as biochemical laboratory techniques or thermodynamic analyses in engineering. The final component is the one we started this chapter with: the scientific *mindset*.

You are expected to acquire these three components in different ways. The knowledge part mostly comes by reading books and scientific papers. Skills are often acquired by "on-the-job" training, in a laboratory or in the field. We frequently expect the mindset to be transferred, as it were, by molecular diffusion or some similar process. The hope is that you will crack the code just by being around researchers in a professional setting. Remarkably, this actually seems to work in many cases, but one purpose of this book is to make the process easier.

Figure 1.1 aims to illustrate that researchers use these three competencies in parallel. It also shows that they can be divided into competencies that are specific to a certain field of research and more general ones. Understanding the scientific approach is a general competency that should be shared by all researchers. The same is true of skills in applying general statistical methods, understanding how to plan and conduct experiments, knowing how to interpret data and so on. It is too common that the research education focuses on specific competencies, like subject matter knowledge and specific laboratory techniques, while neglecting the general aspects. This is probably because they are more difficult to define than the specific ones. Another reason is that the research system is designed in a way that does not naturally engage these competencies in Ph.D. students. Senior researchers tend to plan projects and students tend to work in the laboratory producing data. The aim of this book is to fill this gap in the research education.

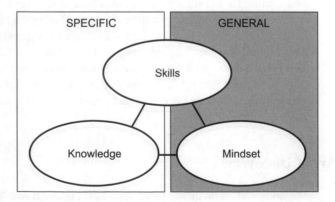

Figure 1.1
Schematic representation of the three components of research competency. It is common that research education focuses on competencies that are specific to a certain field, while neglecting general research competencies.

1.4   About this Book

This book grew out of a need to teach general research competencies. As a Ph.D. student you need to understand how research differs from other activities. If you are working in an experimental field, you need to understand how experimentation differs from other forms of scientific investigation. It is also important to understand the general aspects of collecting and analyzing data. Finally, it is central to understand the scientific process: how to identify research problems, how to approach them, and how to communicate the results in the scientific community. There are good books that cover at least some of these aspects but it would be useful to introduce them in parallel, I thought, between the same pair of covers. It is often difficult for students to fit the pieces together after taking separate courses in the philosophy of science and statistics, for example, especially when the teachers are not experimental researchers themselves. Of course, a single book cannot provide a comprehensive treatment – especially not one as brief as this – but it can serve as a starting point.

The book is organized in two parts. The first part explains different aspects of scientific thinking and the second addresses general methods and techniques that are important to experimenters. The second part often refers to the first, but they can in principle be read separately. It may even be a good idea to skip back and forth between them. Whatever works for you will be fine.

The first part is called "Understanding the World". It contains five chapters, of which you are halfway through the first. The next chapter introduces some central concepts from the philosophy of natural science. It explains that different approaches to scientific exploration result in different types of knowledge, and discusses the connection between observation and theory in science. Chapter 3 very briefly describes how science was invented and how it evolved into its modern form. There is also a discussion about how science has impacted on our civilization. After this, Chapter 4 takes a first look at experiments. It uses Galileo's pioneering experiment on the inclined plane to discuss some interesting aspects of the experimental method. The final chapter of the first part aims to demarcate between research and development, as this is an important distinction for students in applied research fields. There is a discussion about what we mean by theories and how they are developed. It also touches upon creativity, an important aspect of research that is often neglected. Discussing science and experimentation from different perspectives, this part of the book aims to give a better understanding of their essential characteristics.

The second part is called "Interfering With the World" as this is the very basis of the experimental method. It begins with a discussion about what distinguishes experimentation from passive observation. An important message is that observational studies only reveal correlations, whereas experiments can provide evidence for causation. It also explains under which conditions experiments may provide explanatory knowledge. After this, Chapters 7 and 8 cover statistical concepts and techniques that are useful in the collection and analysis of data. Together with Chapter 6 they form a basis for experimental design, which is introduced in Chapter 9. It combines experimental method with statistical techniques to make data collection a more economical and precise process.

The final three chapters describe an experimental research methodology. The methodology draws upon previous parts of the book and is built around a few simple tools. The tools aim to familiarize the reader with ways of thinking that are useful throughout the various stages of a research project. As illustrated in Figure 1.2, this book divides the research process into three phases. During the planning phase, a research question is formulated

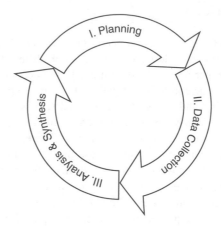

Figure 1.2
The three phases of research.

and an approach is developed to answer it. This is followed by the data collection phase, where a measurement system is devised, analyzed and improved to ensure that the data quality is sufficient for the purpose. After this, a data collection plan is formulated and the actual data are collected. In the third phase, focusing on analysis and synthesis, the data are processed and investigated using graphical and mathematical tools. The aim is to formulate conclusions supported by the data. The synthesis consists of relating these findings to the greater body of scientific knowledge and defining your contribution. The research process is depicted as circular since one experiment often uncovers new aspects of a problem and becomes the starting point of a new investigation.

This research process may not completely cover all branches of experimental research. It may emphasize parts that are less important in some cases, and may even lack features that may be important in others. It is important to note that the methodology is meant to be a pedagogical tool rather than a strict description of how research is done. The purpose is not to provide a cookbook, but to help you organize and elaborate your thinking about how experiments are planned, implemented and interpreted. To put the methodology into a practical context and make the discussion more concrete, we are going to follow two real world experiments through the three phases.

1.5 How to Use this Book

You will find exercises scattered throughout the chapters. The idea is that these should be completed when encountered to ensure that you have understood central concepts and ideas before continuing. There are more exercises in the second part of the book as it covers more technical topics than the first.

I have tried to write in a style that makes it possible to follow the book on one's own. This is also the reason why it is relatively light on mathematics. The amount of mathematics taught at the undergraduate level varies greatly between different fields, so some readers will be less familiar with mathematical concepts than others. Since it is not an end in itself to have to struggle with these concepts, I have tried to be inclusive rather than exclusive in the way that these are presented. This probably means that students in the engineering and physical sciences will be slightly bored by some passages. If you feel that they are not helpful, you should of course feel free to skip over them.

Although most chapters should be quite straightforward to read, some parts are necessarily more technical than others. When reading Chapters 7 through 9 you will probably need to switch to a lower gear and possibly re-read some passages to get through. Do not lose heart over this. It is not your fault that some parts of the terrain are rougher than others.

Many of the computational exercises in this book refer to worksheet functions in Microsoft Excel®. The reason is that this software is very widely available. You can of course complete these exercises just as well in other programs. Some readers will have access to MathWorks MATLAB®, for example, which provides a richer selection of functions and possibilities to write more complex programs.

If you are a Ph.D. student you will probably read this book at the beginning of your studies. It is my hope that it will provide useful ideas and methods that, in time, will be transferred into real skills. But this will not happen unless you apply what you learn to your own research. Reading conveys a theoretical picture but it is only through application that you challenge your understanding and reflect on what the theory really means. For this reason, I hope that this book will become something of a companion during your research studies – a book to return to from time to time when you develop your research ideas.

For those who are using this book to teach a course, a few practical recommendations may be useful. I have found that it helps to make the lectures as varied as possible. It is easier to maintain the audience's attention if you mix statistics lectures with material about scientific thinking, and mix exercises into the lectures. Since the aim of this material is to develop skills rather than knowledge, it is useful to teach and assess the learning outcomes in a practical context. My own approach has been to introduce much of the theoretical material in the first half of a course and devote the second half to simple "research" projects. During these, the class is divided into teams that choose simple research problems where they can apply the methodology and techniques. They have chosen to work with problems like brewing coffee, the rolling resistance of Lego vehicles, the lifetime of soap bubbles, or even strategies for minimizing the time needed to get out of bed in the morning. They begin by identifying interesting aspects of their problems. Is it, for example, possible to control the bitterness of coffee independently of the flavor strength? Or how do stress, comfort and other factors affect how quickly you get up in the morning? The purpose of these discussions is to generate hypotheses that can be tested in experiments. Apart from developing suitable measurement systems and test procedures for such problems, the main challenge is often to develop experiments that allow the teams to *explain* the outcomes, rather than just describe them. I have found that it is important to keep the problems simple enough to let the teams focus on the methodology, rather than to spend undue time getting sophisticated equipment to work. After the analysis they summarize their findings in a "scientific paper", which is peer reviewed by another team before being presented to the others in the form of a conference paper.

Although these problems have little in common with real research problems, they allow the students to apply general research skills to solve real, unstructured problems. They also open up for creative discussions around all the various aspects of solving a research problem. According to the course evaluations, most students perceive these projects as very valuable for turning the theoretical knowledge into practical research skills.

But now it is time to be on our way. There are many stops on our journey and the first is the fundamental question: "What is science?"

Further Reading

Phillips and Pugh [1] provide an interesting and valuable overview of the Ph.D. process.

References

1. Phillips, E.M. and Pugh, D.S. (1994) *How to Get A PhD: A Handbook For Students and Their Supervisors*, 2nd edn, Open University Press, Buckingham.

2 What is Science?

The key to the approach is to keep firmly in mind that the classic position of a researcher is not that of one who knows the right answers but of one who is struggling to find out what the right questions might be!

— E.M. Phillips and D.S. Pugh

The purpose of this chapter is to discuss what science is, and what it is not. This is more of a tall order than it first seems, since science spans so many disciplines. Still, if we are going to learn how to develop science, it is important that we at least reflect on what that means.

Scientists at the beginning of their careers are introduced to the methods of science in different ways. Most learn the tools of the trade from experienced researchers but some complement this training with a course in the philosophy of natural science. Philosophy and science are different subject areas, so philosophers tend to see science from a different perspective than scientists do. You could say that they are more interested in the nature of knowledge than in knowledge of nature. Both perspectives are important but it can sometimes be difficult for science students to merge them in a constructive way. This chapter, along with other sections of this book, is an attempt to bridge the gap between some basic ideas in the philosophy of science and everyday scientific praxis.

Since philosophers and scientists approach science differently, I find it necessary to discuss the methods of science from different standpoints. To avoid confusion, therefore, I will try to indicate when I am speaking from the point of view of a philosopher and when from that of a scientist. I will start by introducing and discussing some basic concepts in the philosophy of natural science. It is important to point out that I neither aim to, nor can, give a deeper account of this subject. This is best left to real philosophers; further reading is suggested at the end of the chapter for those who wish to plunge deeper into it. Later in this chapter I will criticize some of these ideas from the more practical standpoint of a scientist. Although the philosophy of science is important for understanding the nature of scientific knowledge it can be perceived to be detached from the diverse reality of research. I hope that this approach will provide a balanced view in the end.

2.1 Characteristics of the Scientific Approach

Imagine that you are driving a rental car in a foreign country. You have never driven the model of car before and, despite the car being brand new, you find that it does not seem to work properly. Sometimes when you turn the ignition key the engine just will not start. Although you are not a specialist you do have a basic understanding of how an engine works and, starting from there, you begin to investigate the problem.

Experiment!: Planning, Implementing and Interpreting, First Edition. Öivind Andersson.
© 2012 John Wiley & Sons, Ltd. Published 2012 by John Wiley & Sons, Ltd.

Based on your limited knowledge of engines you make a list of potential causes of the problem. Comparing the symptoms you would expect from these causes with your experience of the problem, you find yourself forced to discard one point after another on the list until there are no potential causes left. The next time the engine fails to start you are faced with the fact that you are completely clueless about what to do. In an act of desperation you decide to walk around the car before turning the ignition key again and, to your immense surprise, the engine now starts without a problem. Encouraged, you begin to experiment with this new method and find that walking in a clockwise direction around the car does not work. After a walk in a counter-clockwise direction, however, the engine always starts perfectly. So, in the course of your systematic investigation of the problem you have made a discovery, and a highly unexpected discovery at that!

Later, when returning the car to the rental car office, you complain about the problem. The woman behind the desk asks you if you remembered to push down the brake pedal when turning the key. She explains that this is a safety feature in some cars with automatic gearboxes. To decrease the risk of the car moving during starting, the brake must be pushed down. Thinking back you realize that you must unconsciously have put your foot on the brake pedal when entering the car after the counter-clockwise walk but not after the clockwise walk. That could explain why your method worked.

This may sound like a contrived example but I assure you that it is a true story from life, once told to me by a friend. (Being an engineer he was not particularly proud of trying to walk counter-clockwise around the car, but he admitted to being desperate when doing it.) We will return to this example later. It is useful here because the two methods presented in it could be said to represent two types of knowledge of a problem. Surely, many of us would agree that walking counter-clockwise around the car seems like a less "scientific" method than pushing on the brake. A possible justification is that the latter method is based on a deeper understanding of how cars work. On the other hand, the former method was discovered through a more or less structured investigation of the problem. Isn't that how scientists work? At any rate, we have seen that the fact that a method seems unscientific does not necessarily mean that it does not work.

Scientific research is about obtaining new knowledge, but what kind of knowledge becomes science and how is it obtained? In school, many of us have completed tasks that were called research. They generally involved a visit to the school library to collect information on a topic and summarizing this information in writing. Being able to find information is an important skill for researchers, and scientists do spend considerable time studying their literature, but simply collecting information from books is not research. Since the aim is to acquire new knowledge, it requires something beyond moving facts from one place to another, structuring them neatly and referencing the sources.

Science is often connected with measurements. Are measurements the crucial difference between science and non-science? Measurements are made in different parts of society, from laboratories to the fruit market. In applied research fields, academic researchers and development engineers in industry often make the same kind of measurements. How can it be that a Ph.D. student gets a fancy academic title after using a measurement technique for a few years at a university, while a young engineer making the same type of measurements in industry would be lucky to get a pat on the back for a job well done? If we are not to blame this difference on old tradition, and thereby deprive the Ph.D. degree of its value as an academic merit, there must be a central difference between what the engineer and the scientist do. Something beyond the everyday activities (like calibrating instruments, meticulously following procedures to assure good data quality and taking measurements) in the laboratory. To begin to cast some light over this difference I am going to borrow the following example from Molander [1]:

The main character of the example, Mr. Green, is a keen gardener who one day decides to count the apples in his garden. He goes about the task systematically and methodically and finds out that there are 1493 apples. This is definitely new knowledge, previously unknown to humanity, but is it science? Most people that I ask agree that it is not, even though they may have difficulties explaining why they think so. Those who have published their results in scientific journals and know how results are scrutinized in peer review processes may say that Mr. Green's chances of getting his results published are very slim. But why? It is not because Mr. Green does not have the proper scientific training – even with a Ph.D. in plant physiology he would not get these results published. For his results to become scientifically interesting they need to be incorporated into a greater context, a theoretical framework that gives them generality and helps us better understand some aspect of the world. If Mr. Green had counted his apples every year while also recording information about temperature, precipitation and hours of sun per day he could have searched for relationships in the data. That would approach a scientific way of obtaining new knowledge. When we judge scientific quality it also involves appraising the value of the new knowledge. What is it worth to know something about the number of apples in Mr. Green's garden? Is the garden unique in some sense, for example regarding microclimate, soil, or the type of apples grown there?

Collecting data is an important aspect of research, but it is also important in technical development, politics and other activities that are not considered to be science. There is a wealth of things that we do not yet know and that we could find out, like the oxygen content of the water in a particular bay, or the number of flowers in a particular field. Finding these things out requires planning, meticulous data collection and possibly statistical analysis to provide an unbiased picture of reality. Still, when finding them out we are only describing what we see, we are answering "what" questions. Science goes beyond pure description. It aspires to explain what happens in the world and to predict what will happen under certain circumstances. In other words, it aspires to answer what Phillips and Pugh [2] call "why" questions. Why is the oxygen level low in the bay? Why are there fewer flowers in the field some years? Answering such questions requires something more than careful collection of information. It requires a scientific approach. If you are an engineer working towards a strict deadline in a product development project, or a politician dealing with a problem that suddenly attracts media attention, your need to act can often be more pressing than your need to understand. For the scientist it is always the other way around. In research, trying to find a solution before you understand the nature of the problem is a bit like tying your shoelaces before you put your shoes on.

This means that there are two central and intimately interconnected aspects of science: one that has to do with investigating the world and one that is about interpreting what we see. By investigation we hope to acquire hard facts about reality, and we hope to obtain understanding by interpreting these facts. Interpretation is done within a theoretical framework that allows us to explain the facts. Philosophy books about scientific method often use the words observation and theory instead of investigation and interpretation. Observation sounds more passive than investigation and certainly has a narrower meaning. This is perhaps significant of the fact that such books often focus on how theories are developed. Both aspects are, however, two sides of a coin.

The next few sections are about two basic approaches to research that are described in the philosophy of science. They are written from a philosophical point of view. Even when

I criticize the ideas I speak with the voice of a philosopher. In the remaining parts of this chapter I will again take a scientist's point of view by being a little bit more practically oriented, recognizing the fundamental role of observation in science. I hope to show that the philosophical concepts are useful for understanding the nature of scientific knowledge, although they do not cover all aspects of practical research. We can learn important things from these ideas but should not be too worried if some research does not fit perfectly into their framework.

2.2   The Inductive Method

It is a popular notion that scientists begin by collecting facts through organized observation and then, somehow, derive theories from them. In logic, going from a large set of specific observations to a general, theoretical conclusion in this manner is called *induction* and the approach is, therefore, often called the inductive method. Logic is a branch of philosophy that dates back to Aristotle. It deals with how arguments are made and how to determine if they are true or false. This brief description of the inductive approach follows closely to Chalmers [3], whose book is one of the most widely read introductory books about the philosophy of science. It is useful to describe his rather extreme form of "naive inductivism" in order to highlight some characteristics of the approach, especially those that are generally considered to be its weaknesses.

The inductivist version of science begins with observations, carefully recorded in the form of observation statements. These are always singular statements, meaning that they refer to something that was observed at a particular place and time and in a particular situation. For example, an astronomer might state that the planet Mars was observed at a certain position in the sky at a certain time. The rental car customer from the beginning of this chapter might state that, on a particular day, the engine started only after walking counter-clockwise around the car. To be able to explain some aspect of the world researchers must make generalizations from such singular statements to obtain universal ones, for example that all planets move in elliptic orbits around the sun, or that cars with automatic transmissions require the brake pedal to be pushed down in order to start. The inductivist maintains that it is legitimate to make generalizations from singular observations granted a few conditions are met. Firstly, the number of observations must be large. Secondly, they must be made under a wide variety of conditions. Finally, no observation can be in conflict with the conclusion drawn.

As mentioned above, inductive reasoning moves from a set of specific observations to a general conclusion. You may, for example, note that you become wet when you jump into the water. From a large set of such observations, made under varying conditions, you may infer the general conclusion that water always makes you wet. Going the opposite way, from general statements to specific conclusions, is called *deduction*. Consider the following example of deductive logic:

Premise 1:   All scientists are mad.
Premise 2:   I am a scientist.
Conclusion: I am mad.

Here, the conclusion is a necessary logical consequence of the two premises. Premise 1 may be held to be a general law, derived from a large set of observations by induction. The problem is that the truth of the conclusion depends on the premises being true, which they

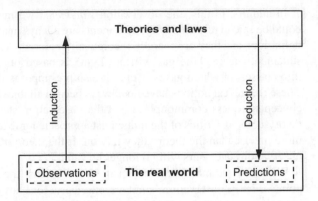

Figure 2.1
Induction moves from a set of specific observations to a general conclusion. Deduction moves from general statements to a set of expected, specific observations.

obviously (at least to me) are not in this case. The weak link is induction because it cannot be logically justified. Even if you have made a very large number of consistent observations under various conditions, it does not follow that every future observation will be consistent with them. They provide some degree of support for your conclusion, but do not prove it to be true. The principles of induction and deduction are described graphically in Figure 2.1.

Chalmers uses a moral tale to drive this point home in an elaboration of Bertrand Russell's story of the inductivist turkey [2]:

A turkey found that, on his first morning at the turkey farm, he was fed at 9 a.m. Being a good inductivist, he did not jump to conclusions. He waited until he had collected a large number of observations of the fact that he was fed at 9 a.m., and he made these observations under a wide variety of circumstances: on Wednesdays and Thursdays, on warm days and cold days, on rainy days and dry days. Each day, he added another observation statement to his list. Finally, his inductivist conscience was satisfied and he carried out an inductive inference to conclude, "I am always fed at 9 a.m.". Alas, this conclusion was shown to be false in no uncertain manner when, on Christmas Eve, instead of being fed, he was killed.

So, the inductivist's first condition is problematic: however large the number observations they can never ensure the truth of the conclusion. Let's look at the next condition and ask what counts as significant variation in circumstances. The turkey in the example made his observations on different weekdays and under varying weather conditions but not under sufficient variation of holiday seasons to find out what happens to turkeys on Christmas Eve. Determining the boiling point of water is perhaps an example more relevant to science. What should be changed to fulfill the criterion of sufficiently varying circumstances in that case? Is it necessary to vary the ambient pressure, or the purity of the water? Should we try different methods of heating, or different weekdays for the measurements? Most people would agree that the first two variables, pressure and purity, are sensible candidates, whereas the latter two make less sense. The point is that there are infinite ways to vary the circumstances, so how do we know which ones to choose? Well, we probably have some expectations of what will affect the boiling point of water and these expectations are based on some level of theoretical knowledge about the world. Theory could thereby be claimed to play a role prior to observation. In that case the inductivist assumption that science starts with observation does not hold true.

In addition to this, the observation *statements* we make to describe our observations could be said to rely on theory. Statements are always made in some language that contains definitions and they are thereby theoretically colored. As Chalmers puts it, even a simple statement such as "the gas will not light" is based on a theory that divides substances into classes, of which gases is one. It also presupposes that some gases are combustible. These theoretical notions have not always been available. In modern science the theoretical concepts are less commonplace and the observation statements may be more obviously theory-colored. Critics of the inductivist approach argue that observation statements are no more precise than the theory they rely on. If the theory turns out to be false, they say, the observation statements will no longer be valid.

We could summarize the criticism raised against the inductive method so far in two points. Firstly, observation is not a secure basis for knowledge. This is because induction in itself cannot be logically justified. We may make as many observations of something as we like, we still cannot use them to prove the truth of our conclusion. There is always the possibility that one day we will make an observation that contradicts the conclusion, as our inductivist turkey did on Christmas Eve. Secondly, it has been argued that observations and observation statements involve theory and, therefore, are just as fallible as theories are.

Besides these points there is another important weakness of a purely inductive approach that has to do with the type of knowledge that it produces. The problem is how singular observation statements can be generalized into theories that are scientifically useful. Science aims to explain and predict. Theories without explanatory power are thereby of limited use in science. If we note that we become wet every time we jump in the water we may infer that water makes people wet, but we cannot obtain theories about the underlying properties of water and people that explain *why* we get wet. The astronomer who observes the position of the planet Mars in the sky is perhaps a more illuminating example – how does the inductivist generalize from such observations to obtain Kepler's first law, that planets move in elliptic orbits with the sun as a focus? At the beginning of this chapter I said that the two solutions to the problem with the rental car represented different types of knowledge. The customer knew that walking around the car worked, at least if it was done in the counter-clockwise direction, but was not able to explain why. This is the type of knowledge that an inductive approach can provide us with. The woman at the rental car office knew that pushing down the brake pedal worked but she could also explain that this was a result of a certain safety feature of the particular car model. Can you think of any way to obtain this knowledge by a purely inductive approach?

Humankind has long noticed that there are regularities in Nature. Things tend to happen in certain ways and not in others, as if they were guided by machinery hidden from our eyes. The ultimate aim of science is to explain, rather than just describe, what happens in the world. As researchers we must therefore aspire to understand something of that hidden machinery. Since it is hidden from our eyes this task requires a more imaginative approach than just watching what goes on in the world and making notes of it.

2.3   The Hypothetico-Deductive Method

When the young Isaac Newton was toying with prisms in 1666 he noticed that they dispersed a white beam of sunlight into all the colors of the rainbow. If the colored light from a part of this rainbow was passed through a second prism it did not result in a new rainbow, only in further dispersion of the colored beam. The experiment is described schematically in

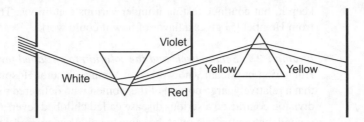

Figure 2.2
Schematic representation of Newton's experimentum crucis. The second prism does not produce rainbow colors, as the first one does. Newton concluded that prisms do not add color to white light, but the colors are constituents of white light.

Figure 2.2. Newton concluded that the prism did not add the rainbow colors to the light. Maybe the white light of a sunbeam already contained all the colors and all the prism did was to spread them out over the wall opposite his window? As a consequence, he thought, re-combining the colors should produce white light [4]. He set up an experiment to do just this and, as most readers are probably aware, his thought was confirmed.

This is an example of working out the consequence of an hypothesis by logical deduction and testing it by an experiment (although Newton himself would not agree that this is the full story of what he did [4]). An hypothesis can be seen as a preliminary attempt to explain something – a theoretical statement without much observational support. We shall now discuss a method based on this procedure, where scientific inquiry begins with theory instead of observation. According to it, a scientist faced with a seemingly inexplicable phenomenon should begin by proposing speculative theories, or hypotheses, about its cause. These can then be tested rigorously against observation, for example by controlled experiments, to see if the consequences that follow from them correspond with reality. Working out consequences from premises is called deduction, speculative theories are called hypotheses and, accordingly, the method is called hypothetico-deductive. My description of the method closely follows how Chalmers describes falsificationism [3], which is a variant of the method proposed by Karl Popper. Most working scientists seem to apply some of the method's central elements, not least since they often discuss hypotheses and generally agree that hypotheses must be testable in order to be scientifically useful. In the words of Popper, hypotheses must be falsifiable.

If an hypothesis is formulated in a way that makes it theoretically impossible to disprove, it is not falsifiable. For example, a recruitment firm that throws half of the applications for a job into the wastebasket with the motivation "we don't want to recruit unlucky people" bases its decision on a non-falsifiable hypothesis. It is non-falsifiable because people whose applications end up in the wastebasket are, by definition, unlucky. Why would their applications otherwise be thrown away? It is a bit like saying that all sides of an equilateral triangle have equal length. If they did not it would not be an equilateral triangle – the statement is logically impossible to contradict. Non-falsifiable hypotheses are statements that are true whatever properties the world may have. Such hypotheses tell us nothing about the world and are, therefore, of no use in science.

According to the hypothetico-deductivist, research begins with an hypothesis and proceeds by testing it against observation. If our observations contradict the hypothesis we discard it and look for a new one. If the hypothesis is supported by our observations we keep it, but continue testing it under various conditions. The following classical example from Hempel [5] gives a flavor of how it could work.

Example 2.1: Semmelweis and the solution to childbed fever Ignaz Semmelweis was a Hungarian physician who worked at Vienna General Hospital in the mid-1800s. He noted that a relatively large portion of the women who delivered their babies at the first maternity division contracted a serious disease called childbed fever. About 7–11% died from it. The second maternity division at the same hospital accommodated about as many women as the first, but only 2–3% of them died from childbed fever. The theory that diseases are caused by microorganisms would not be widely accepted until the late 1800s, so Semmelweis considered various other explanations. He quickly dismissed some of them as they were contradicted by what he already knew. One of these was the then common idea that diseases were caused by "epidemic influences", attributed to "atmospheric-cosmic-telluric changes", affecting whole districts. If this were so, how could the first division be plagued by the disease while the second was relatively spared? He also noted that women who gave birth in the street had a much lower risk of contracting childbed fever, although they were later admitted to the first division. These were often women who lived far from the hospital and were not able to arrive in time after going into labor. A street birth just outside the hospital should not decrease the risk for epidemic influences.

Could the problem be due to overcrowding? Semmelweis saw that it was not. The second division tended to be more crowded, partly because the patients desperately wanted to avoid the notorious first division. Instead, he turned to psychological factors: The priest, bearing the last sacrament to dying women, had to pass five wards to reach the sick-room of the first division. He was accompanied by an assistant ringing a bell. Presumably, this display would be terrifying to the other patients. Fear could explain the fever, because in the second division the priest could access the sick-room directly, without passing other patients. To test his idea Semmelweis persuaded the priest to take a roundabout route and enter the sick-room unobserved and silently, without ringing the bell. Unfortunately, this had no effect on the mortality in the first division.

Semmelweis also noticed another difference. In the first division the women were delivered lying on their backs, while in the second division they lay on their sides during birth. He decided to try the lateral position in the first division but, again, without result.

In 1847 he got a decisive clue to the solution. During an autopsy, a colleague of his received a puncture wound in the finger from a scalpel. Later, the colleague showed the same symptoms as the women that had contracted childbed fever and, eventually, he died. As previously mentioned, the role of microorganisms was not yet recognized but Semmelweis started to suspect that some kind of "cadavic matter" caused the disease. In that case, the physicians and their students would carry infectious material to the wards directly from dissections in the autopsy room. The difference in mortality between the divisions could then be explained by the patients in the second division being attended by to midwives, whose training did not include dissections. To test his hypothesis he required all medical students to wash their hands in a solution of chlorinated lime, which he assumed would destroy the infectious material chemically, before examining the patients. As a result, in 1848 the mortality in childbed fever decreased to 1.27% in the first division. This was fully comparable with the 1.33% in the second division that year. The hypothesis was further supported by the lower mortality among women giving street birth: mothers who arrived with babies in their arms were rarely examined after admission and, thereby, escaped infection. ■

We see how Semmelweis proceeded by first trying to understand the cause of the disease and then comparing his ideas to the information at hand, or to information obtained by specific experiments. Unsuccessful ideas were discarded and new ones proposed until the problem was solved. We may also note that, although his idea about "cadavic matter" improved the understanding of the disease, it was still far from the more modern, microbiological theory. Although it may be tempting for modern readers to exchange the term with "bacteria", Semmelweis did not understand that living organisms caused the disease. It is again clear that observational support does not prove an hypothesis to be true. The hypothetico-deductivist must abandon the claim that theories can be proved by observation, at least in a strictly logical sense. Still, hypotheses that resist falsification in a wide variety of tests tend to become established theories with time, as the support for them grows ever stronger. Even though they are not proven true, we may still acknowledge that they are the best theories available, since they are supported by the most evidence.

Finally, hypotheses should not only be falsifiable, they should even be as falsifiable as possible. It is difficult to contradict a vague statement. For science to progress there cannot be any doubt whether an observation supports an hypothesis or not. A good hypothesis is falsifiable just because it makes precise assertions about the world. It becomes more falsifiable the more wide-ranging the claims it makes, since the number of observations that could potentially contradict it increases. For instance, Newton's theory for the movement of the planets around the sun is more falsifiable than Kepler's older theory. Newton's theory consists of his three laws of motion and the law of gravity, stating that all pairs of objects in the universe attract each other with a force that varies inversely with the square of their separation. Whereas Kepler's laws of planetary motion only apply to the planets, Newton's theory is more general: it describes planetary motion as well as a large number of phenomena, including falling bodies, the motion of pendulums, and the relationship of the tides to the positions of the sun and moon. This means that there are more opportunities to make observations that contradict Newton's theory. In other words, the set of potential falsifiers of Kepler's theory is a subset of the potential falsifiers of Newton's theory, which then becomes more falsifiable.

2.4 Consequences of Falsification

When the scientific knowledge grows within a field, the theories of that field are sometimes affected. New insights are not always compatible with current theory. There are, in principle, three possible scenarios in such cases. The theory may be modified to explain more aspects of the world, it may be completely replaced, or, sometimes, it is not affected at all. We will continue to use the laws of planetary motion as examples to illustrate these scenarios.

Today, few physics undergraduates learn about Ptolemy's theory for planetary motion. This is because it has been proven wrong and has been successively replaced with better theories. Ptolemy lived in the second century of the Common Era (CE) and his theory was based on Aristotle's cosmology, stating that the earth was situated at the center of the universe with the sun and planets revolving around it. The planets moved in perfect circles with uniform motion, since nothing in heaven could deviate from perfection.

Readers who have studied the motions of the planets in the night sky may have noticed a curious effect called retrograde motion. Looking night after night at the planet Mars, for example, it seems to move westward relative to the backdrop of the stars. At some point it suddenly makes a halt and starts to move backwards. Later, the backwards motion ceases and the planet continues in the westward direction. This motion makes perfect sense in a heliocentric system, where the earth revolves with the other planets around the sun. From our point of view it *looks* as if Mars moves backwards, because the earth "overtakes" it in its inner orbit. In the Ptolemaic system this motion was, of course, more difficult to explain. Ptolemy attempted to do so by letting the planets move in smaller circular orbits, called epicycles. These, in turn, revolved in larger, circular orbits around the earth. This solved the problem at the cost of immense complexity. Instead of one circular orbit for each planet the Ptolemaic universe consisted of several dozens of cycles and epicycles rotating upon and about each other. There were equants, eccentrics, deferents, but still this system could only approximately match the observations. By the sixteenth century the calendar and the seasons diverged by weeks, and predictions of eclipses and conjunctions could miss their mark by a month [6]. In the hypothetico-deductive account of science the Ptolemaic system had been falsified since the observations did not match its predictions.

Copernicus realized that a heliocentric system was a more accurate model of the universe. He placed the planets and the earth in orbit around the sun and the moon in orbit around the earth, but held on to the assumption of circular orbits. His system was both more accurate and mathematically simpler than the Ptolemaic one, though it still made use of epicycles to better match astronomical observations. Although modern astronomers do not accept the details of the Copernican system, the heliocentric theory of the solar system has certainly replaced the geocentric one. This is an example of how a theory that fails to describe reality may eventually be replaced.

In the hypothetico-deductive approach, a falsified theory should be discarded and replaced with a new and better theory. The new theory should account for more phenomena than old theory did. Besides explaining the phenomena that falsified the old theory, it should also account for everything that the old theory explained. Then, how can it be that some theories survive falsification? Consider, for instance, Newton's laws of motion. At the beginning of the last century physicists found that these laws, which had been believed to be universal laws of motion, did not apply to the motion of the tiny constituents of atoms. On the scale of the very small, the established theories describing our everyday world broke down completely. This became the starting point of a new theory called quantum mechanics, developed to describe the microscopic world of atoms and elementary particles. Newton's theory had been falsified since it had been shown not to be as general as previously believed. An area had been found in which it did not apply, but should it be discarded?

Completely replacing an established theory is a formidable task. At the beginning of the twentieth century Newton's theory had been supported by a plethora of various observational tests for more than two centuries. It had predicted new phenomena and had even been used to predict the existence of the previously unknown planet Neptune. With such merits it would be preposterous to suggest that the theory was completely false. It had obviously been brought to its limit when it was confronted with the mysteries of the quantum world, but it still applied to the macroscopic world.

The new theory of quantum mechanics is indeed more falsifiable, since it includes Newton's mechanics as a special case, and this is an important aspect of the growth of science. But although quantum mechanics may be a more general theory, there is no doubt that Newtonian mechanics are still used extensively by scientists and engineers to solve problems on the macroscopic scale. It is an example of a theory that has been falsified without being replaced. It is still valid and useful but we are more conscious of its limitations.

As previously mentioned, there is another alternative to replacing a whole theory that is in trouble. A smaller modification might save it from being falsified. Newton's laws have previously been described as an improvement over Kepler's laws. These, in turn, were an improvement on Copernicus' heliocentric theory. Johannes Kepler had access to the best astronomical data available before the invention of the telescope and struggled to find a mathematical representation that fitted with them. After years of calculations he overthrew the last two fundamental assumptions of the Aristotelian cosmology – that planets move in circles and that the motion is uniform. His system can be described as an addition to the heliocentric theory. The planets were ordered around the sun as in the Copernican system but they moved in elliptic orbits with the sun in one focus and they moved faster when they were closer to the sun. Since this addition removed the discrepancy between the data and the heliocentric theory, the theory was saved from falsification. Most readers will probably agree that his addition was a legitimate one but the following example illustrates that not all modifications of theories are.

A central assumption in Aristotle's cosmology was that nothing in the heavens could deviate from perfection. As a result, it was long believed that all celestial objects were perfect spheres. When Galileo Galilei pointed his telescope towards the moon in 1609 he was intrigued to find that it was far from a perfect sphere. It was scattered with mountains, some of which he estimated were more than 6000 meters high [7]. One of his Aristotelian adversaries agreed that it looked that way in the telescope but cunningly suggested that there was an invisible substance filling the valleys and covering the mountains in such a way that the moon was still a perfect sphere. When Galileo asked how this invisible substance could be detected, the answer was that it couldn't. The only purpose of this additional theory was to save the original one. It was an *ad hoc* modification. The moon was still spherical, it just did not look that way. Any modification of a theory must of course be testable. It must also be testable independently of the theory that is saved from falsification. Galileo went right to the core of this problem in his brilliant response to his adversary. He agreed that there was an invisible and undetectable substance on the moon but instead of filling out the valleys he suggested that it was piled on top of the mountains, so that the surface was even less smooth than it appeared in his telescope [3].

2.5 The Role of Confirmation

From now on I am going to switch perspectives. So far I have described science from the point of view of philosophy, which is an outside perspective. Researchers, who apply the scientific method every day, tend to look at it differently. Many scientists think that the hypothetico-deductive method, and falsificationism in particular, gets a negative taint by stating that science progresses by disproving theories. With a little bit of travesty, falsificationists would remain unmoved if their observations were to support their hypotheses. Instead of cheering at the confirmation of their ideas they would begin to think about new ways to kill them. This is counter-intuitive to most working scientists. People are seldom awarded Nobel prizes for having disproved a theory but rather for discovering and explaining new phenomena. In all fairness, confirmation does play a role in the hypothetico-deductive approach [3]. If a bold hypothesis is confirmed it corresponds to the discovery of something unlikely. Similarly, if a cautious hypothesis is falsified it means that something that seems self-evident is, in fact, incorrect. Both of these scenarios represent significant advances in science. Conversely, little is learnt from the falsification of a bold theory (something unlikely is false) or confirmation of a cautious theory (something likely is true). For the hypothetico-deductivist, confirmation is important so long as it is of novel predictions resulting from bold hypotheses [3].

Problems do arise with the hypothetico-deductive method when we realize that observations that are contrary to a theory do not necessarily render it false. This was arguably the case with Newton's mechanics. I think that the following (imaginary) example shows that scientists do not necessarily think like falsificationists. Imagine that we discover that a certain type of mold kills bacteria. We may then hypothesize that it can be used as an antidote to bacterial infection. To test the hypothesis we set up a program of double blind tests on a group and a control group of patients. The results of the study turn out to be encouraging and, after follow-up studies, we start using the drug and successfully treat patients on a large scale. We may now ask ourselves if we have proven our hypothesis. The falsificationist would firmly say no; it is not logically possible to prove an hypothesis by observation. But many scientists would consider the statement of the hypothesis to be true, at least until proven otherwise. Most scientists are more pragmatic than many philosophers of science and tend to consider theories to be "innocent until proven guilty".

Now, to add some excitement to this research program, suppose that some bacteria develop a resistance to the drug. As a result, some patients do not recover when treated with it. Does this new state of affairs disprove the theory? The falsificationist would probably vote yes, although I think it is evident from the example that it is difficult to make a clear-cut decision. Large-scale treatment of patients with a new drug changes the environment in which the bacteria live. This exerts an evolutionary pressure on the bacteria to adapt to the new situation. If a mutation in a bacterium improves its ability to survive in the new environment, its genes will gradually become more common in the gene pool and a resistant strain will develop. The state of Nature has thus changed after the study was made. Figure 2.3 illustrates these events and how a falsificationist would react to them.

As scientists, we may prefer to make an addition to our theory instead of discarding it altogether. We may say that the drug works as an antidote to bacterial infection, *except* when the bacteria have developed resistance to the drug. Does it sound like an *ad hoc* modification? Well, if we can think of a way to identify the resistant bacteria without using the drug the modification is independently testable. Otherwise it is not. So, if the addition is to be considered to be *ad hoc* or not depends on if we have enough knowledge to test it independently of the original theory. If that knowledge is not in place, the falsificationist would want us to discard the theory. In reality, scientists would probably know better than that. After all, the drug used to be completely effective. The fact that it has become less effective will certainly inspire us to look deeper into the problem but not to discard a theory that has been confirmed in rigorous tests. The trustworthiness of theories is not quite so black and white as the falsificationist describes it to be. There is often a gradual transition from the suspicion that there is a problem to the realization that the theory no longer fills its purpose; practically minded people will probably realize when it is time to abandon the theory without consulting the philosophy books.

I ended the section about inductivism by saying that scientists must strive to understand the hidden machinery of Nature to be able to explain, rather than just describe, what they see. Since the hypothetico-deductive method begins by formulating hypotheses, the striving to understand this machinery is built into the method. As opposed to pure inductivism, it has the ability to produce theories with *explanatory* power. Although the example above illustrates some of the risks of following it too slavishly, the basic principle of the hypothetico-deductive method is highly useful. This is why the remaining parts of this book tend to

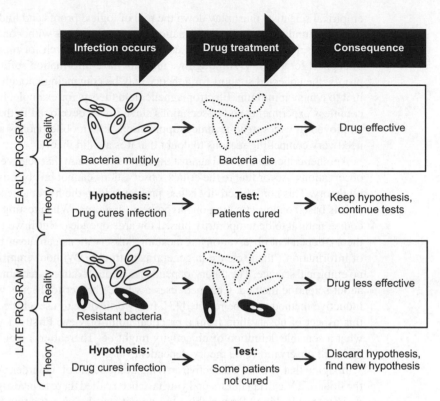

Figure 2.3
An imaginary example of testing an hypothesis. In the first trials the drug kills the bacteria and the patients are cured. The hypothesis is kept and subjected to further tests. At a later time, a resistant strain has developed. The drug is suddenly less effective and the hypothesis should be discarded. But has it been disproved?

favor it over the inductive method. Before we discard inductivism completely, however, let us discuss if the criticism against it really holds together from a scientist's perspective.

2.6   Perception is Personal

According to the inductive method, scientists seek – and find – the truth without being guided by personal opinions or preferences. Inductive science begins with objective observation. The resulting theories always enter the picture at a later stage.

One of the criticisms against this method that we have mentioned is that theories cannot be proved on the basis of observations. This problem remains unresolved in the hypothetico-deductive method. If we define truth in strictly logical terms, we have to live with a certain degree of uncertainty about the truth of our theories, regardless of which method we use. To prove something logically, we need access to all the facts relevant to our problem. This is possible, for example, in mathematics, where theories are derived from axioms laid down by people. Axioms can be said to define the researcher's private universe, because if we make the rules ourselves we have access to a complete set of facts. In the natural sciences, Mother Nature makes the rules and the researcher's job is to find them. For this reason, empirical scientists must play down the role of logical proofs and find ways to compensate for the naturally incomplete information they have to work with – and they have to do this whatever scientific approach they use. One way is to search for many independent types of support for a theory. Although we may not have information sufficient for valid logical proofs, the collected support for a theory may be convincing enough to reduce our doubts in it to a mere minimum. Electromagnetic field theory, for example, has been supported by countless experiments and observations since it was developed in the 1800s and has been employed in widespread technical applications like mobile phones and radar. You would need very compelling reasons to doubt that it is a valid theory.

Another criticism is raised against the inductivist claim that science is based on objective observations. According to the critics, observations cannot be objective as they are colored by theory. This non-objectivity is less problematic in the hypothetico-deductive approach, as it is based on investigating one hypothesis at a time. When testing an hypothesis it is of course natural to be temporarily biased towards one idea. But if we understand the talk of theory-dependent observations as meaning "observations can never be trusted as a source of information", then science in general is in trouble. Without empirical data there can be no empirical science. The theory-dependence of observation is a main theme in many books about scientific method. Chalmers uses a whole chapter to discuss why it means that the inductive method must be rejected [3]. It is worthwhile to use a little space here to discuss this aspect of observation from a practical point of view. Firstly, I would like to discuss what a sensible definition of objectivity might be. Thereafter, I would like to ask in what sense all observations are theory-dependent.

Imagine that you are a detective arriving at the scene of a murder. A lifeless body lies on the floor of a Victorian library and you have been called there to investigate the crime. Where do you start looking? Presumably, you do not start by investigating the wood paneling on the walls to see if it is made from oak or teak. Nor do you try to determine the optical quality of the glass in the crystal chandelier. No, you probably start by looking at the body. How is it positioned? Are there any indications of violence? It is likely that you continue with particular aspects of the rest of the scene. Did the murderer leave any traces? Is something missing that could provide a clue about the motive? There is an immense wealth of facts that could be observed in the library and to solve the case you obviously have to *choose* what to observe. Strictly speaking, this means that you are not objective. Still, most people would not accuse you of being biased if you conducted your investigation this way, because they think that objectivity has to do with having an open mind and not jumping to conclusions before looking at the relevant facts. This is the common sense definition of objectivity. Figure 2.4 illustrates how it differs from the strict definition.

Since our knowledge is incomplete (otherwise we would not need to investigate the murder) it is difficult to determine beforehand which facts are the most relevant, but the examples above show that many facts can be discarded as irrelevant at an early stage. If later findings should suggest that the murderer was looking for diamonds and that these may have been hidden among the crystals of the chandelier, a closer look at the chandelier will be justified. But before we know anything about the motive it would be foolish to pay the chandelier special attention. If objectivity means giving the same weight to every single fact, an objective investigation is probably impossible and, at any rate, it could never solve the case. This strict definition of objectivity is impractical. Regardless of if we are solving a scientific problem or if we are making a cheese sandwich we must prioritize between observations, concentrate on the important ones and disregard the unimportant ones. It is

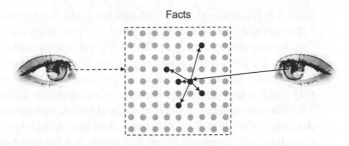

Figure 2.4
The observer on the left is strictly objective and pays equal attention to all facts. The observer on the right selectively pays attention to facts relevant to the problem. The larger the number of facts, the less successful the left observer is at solving the problem.

not sensible to say that an observer who does not pay equal attention to all facts is not objective.

Before we leave our detective to go about his/her investigation at the scene of this hideous crime we may stop and ask *how* he/she knows what to observe and what to ignore. He/she probably has an idea about how murders happen and how such events leave traces on a crime scene. This knowledge comes from a mixture of training, experience and common sense. We can think of such knowledge as a filter placed between the facts of the scene and the mind of the investigator. It filters out irrelevant information and transmits the aspects that should be prioritized. A good filter clearly distinguishes between the relevant and irrelevant facts, while a poor one is less reliable. The quality of the filter could be called the investigator's power of observation and it can be improved with training, experience and reflection. This means that not all observers observe the same things. Even a single observer may observe different things on different occasions if the filter should change over time. Acknowledging that observation is a skill does not necessarily imply that skilled observers are less objective than others – at least not if we use the common sense definition of objectivity. It could simply mean that skilled observers are more efficient at getting an objective view of the problem under study. We should not empty our heads of all knowledge every time we try to solve a problem. At least not if we want to solve it.

Let's turn to the other question. In what sense are all observations theory-dependent? What observers see depends on their knowledge, skill and personality. Perception, in other words, is personal. Chalmers uses several examples to underline this [3]. One of them is how a student and an expert radiologist perceive X-ray images. Where the student may only see shadows of the heart and ribs, with a few spidery blotches between them, the expert sees a wealth of significant features. With training, the student may gradually forget about the ribs and begin to see the lungs as well as a rich panorama of details on them. Another example from Chalmers is an amusing entry from Johannes Kepler's notebook, made after observation through a Galilean telescope. It reads, "Mars is square and intensely coloured". This statement can be said to have relied on the theory that the telescope gave a true picture of reality, which it clearly did not. Chalmers proceeds to say that this means that "observation statements do not constitute a firm basis on which scientific knowledge can be founded, because they are fallible".

The point I wish to make is that the fact that observations, and observation statements, are fallible does not undermine their role as fundamental information carriers in science. I think Chalmers would agree with this but I still want to make the point because it is easy to be overwhelmed by the alleged problems with observation when reading statements like this. It is worth mentioning again that if observations could not be trusted there could be no science. We know from a vast body of experience that science is, in fact, a highly successful endeavor. Of course, observers can make mistakes when observing or when explaining what they see, just as philosophers can make mistakes while thinking or expressing their thoughts. That does not undermine the fundamental role of thoughts in philosophy. When discussing the theory-dependence of observation it is easy to indulge in pessimistic statements about how unreliable observations are and forget that the proof of the pudding is, in fact, in the eating. If faulty observations lead to a theory that is incorrect, this will not go unnoticed. Observations must be repeated and elaborated independently by several researchers to have an impact. Scientists do not necessarily make fewer mistakes than other people, neither are they less prone to be led astray by their pet ideas, but the methods of science have elements built into them that make it more difficult for scientists to fool themselves. For observations to have scientific impact we demand a high degree of inter-observer correlation. This means that many observers must be able to see the same thing and that it must be possible for other researchers to successfully repeat our experiments. Before publication, scientific results are examined and criticized by other researchers in anonymous peer review processes. These systems are in place because we are aware of the risk of mistakes, and to increase the degree of objectivity in the results.

To say that all observations are theory-dependent is, in practice, misleading. Some observations are indeed heavily theory-dependent, while others are not. For instance, if our detective remarks that the victim lies face-down on the floor, this statement does not presuppose any theoretical knowledge at all. Stating that the victim was beaten with a blunt instrument only presupposes that such instruments leave other types of traces on a body than sharp ones. To call this knowledge a theory empties the word of any meaning relevant in science. If, on the other hand, scientists say that they have been observing a single electron oscillating in the electric field of a laser beam, this observation statement clearly relies on involved, high-level theory. If this theory turns out to be false it could render the observation statement invalid but, as we soon shall see, this is not necessarily so.

For those who think that observation statements are always interpretations in the light of theories, Hacking provides ample examples of the opposite [8]. I am going to mention a few of them briefly here. One of them is the discovery of double refraction in Iceland Spar (calcite) by Erasmus Bartholin in 1689. If you put a piece of this material over a printed page you see the text double. Iceland Spar was the first known producer of polarized light. This phenomenon had to wait well over a century for Fresnel's theoretical explanation, so it is clear that the discovery did not rely on theory. Another optical phenomenon that was discovered long before any theoretical explanation is diffraction. Grimaldi, and later Hooke, discovered that there was illumination in the shadow of an opaque body. Careful observation revealed regularly spaced bands at the edge of the shadow, which come from diffraction. These are examples of phenomena discovered by alert observers. What about deliberate experiments then, do not they always involve theoretical assumptions? If we return for a second to the frustrated rental car customer from the beginning of this chapter, we may recall that walking counter-clockwise around the car clearly was an experiment without theoretical motivation. Similar examples are not uncommon in science. One of Hacking's examples is David Brewster, who experimentally determined the laws of reflection and refraction of polarized light, and also managed to induce birefringence (polarizing properties) in materials under stress. These things were later theoretically explained within Fresnel's wave theory but Brewster himself was advocating the Newtonian view that light consisted of rays of corpuscles. As Hacking puts it, "Brewster was not testing or comparing theories at all. He was trying to find out how light behaves." This summarizes the important point that our understanding of Nature's machinery often begins with the discovery of an interesting phenomenon, and not necessarily with theoretical considerations. I could go on quoting Hacking, whose book is full of convincing examples like these. To establish that there are plenty of examples also outside it, and thereby strengthen the point, I would like to add a couple of examples of my own choosing. I will first briefly mention Gregor Mendel, who is known to many as a brilliant experimental biologist. The experiments leading to his discovery of the laws of inheritance are described in a separate example in Chapter 6. Here, I will restrict myself to pointing out that this discovery had to wait half a century to be acknowledged by the scientific community, because there was no theory available that gave them meaning. My other example comes from the history of chemistry. In this connection I cannot resist referring back to the observation statement "the gas will not light", used in the section on inductivism to exemplify a theoretically colored statement. The following example involves similar statements but made in a scientifically more relevant context.

Example 2.2: Joseph Priestley and the discovery of oxygen   Before Antoine Lavoisier introduced the oxygen theory, the phlogiston theory was the prevailing theory of combustion. Phlogiston was believed to be a material that escaped from substances during combustion. It can be thought of as a slightly modernized version of Aristotle's fire element, with the difference that its existence had found some support in experiments. Besides combustion, phlogiston was also believed to be involved in calcination of metals. When metals were heated in air they were believed to change because they lost phlogiston. The resulting calxes (oxides) changed back into metals when heated with charcoal, as they regained phlogiston from it. Since the charcoal only left a small amount of ash after the process it was presumed to be rich in phlogiston. Furthermore, combustion could not be sustained in vacuum since air was needed to absorb phlogiston. Correspondingly, combustion in a sealed container ceased when the air inside it was saturated with phlogiston. The theory thus explained a range of phenomena, but it had a weak point: metal oxides lose weight when they are reduced to metals. How could that be the case when they gained phlogiston? [8]

The gas formed when reducing calxes with charcoal was called fixed air. We know it as carbon dioxide. It had been found that fixed air was produced also during fermentation and respiration and it was known not to support life. In August 1774 the English chemist Joseph Priestley collected a gas that had been formed when heating *Mercurius Calcinatus* (mercury oxide) *without* charcoal. His investigations showed that the gas had properties opposite to those of fixed air: "what surprised me more than I can well express, was that a candle burned in this air with a remarkably vigorous flame" [9]. Later, he noticed that a mouse survived twice as long in the gas as in ordinary air. Encouraged by this result he decided to breathe the gas himself: "I fancied that my breast felt peculiarly light and easy for some time afterwards. Who can tell that but in time, this pure air may become a fashionable article in luxury. Hitherto only two mice and myself have had the privilege of breathing it" [9]. He was breathing pure oxygen but he called it *dephlogisticated air* because, in the phlogiston system, a gas was thought to promote combustion if it was deficient in phlogiston. This gave the phlogiston in the burning material somewhere to go [9].

Antoine Lavoisier, who made more quantitative studies, later discarded the phlogiston theory. He realized that the gas that Priestley had collected was not a variety of air, but a separate component of air with unique properties. It was he who named it oxygen. Since Priestley never stopped supporting the phlogiston theory we know that he interpreted his observations using a false theory. Despite this, his observation statements make perfect sense to us. If his experiments were to be repeated today by someone ignorant of the idea of phlogiston, the results would probably be described very similarly. Clearly, not all observation statements are theoretically colored. Indeed, not all experiments presuppose theory, as Priestley's own reflection on his test with the candle flame shows:

> *I cannot, at this distance of time, recollect what it was that I had in view in making this experiment; but I know I had no expectation of the real issue of it. If [. . .] I had not happened for some other purpose, to have a lighted candle before me, I should probably never had made the trial; and the whole train of my future experiments relating to this kind of air might have been prevented [10].* ∎

There are many examples of scientists who have said that they work from observations to build theories. Charles Darwin, the father of modern biology and one of the most astute intellects in the history of natural science, described himself as "a kind of machine for grinding theories out of huge assemblages of facts" [11]. This sounds like a rather inductivist approach, even though it is probably not a complete description of his way of working. The point is that theories do not pop up from nowhere. They are developed because there is a phenomenon that needs explanation, and the phenomenon must be discovered before a theory can be considered. Once Darwin's great idea had been born, he developed it by testing it against facts collected by him and the many people he corresponded with. His approach probably contained elements from both the inductive and hypothetico-deductive method, and he was not alone in this mode of working. Newton himself said that the best and safest method of research is, "first to inquire diligently into the properties of things, and establishing those properties by experiments and then to proceed more slowly to hypotheses for explanations of them" [4]. I am sure that many with me have experienced, when reviewing data from an experiment, that an unexpected aspect of the data suddenly appears, spurring new ideas that may completely change the course of the investigation. Science does not have to start with a theoretical assumption. Noticing something unexpected that tickles one's curiosity is quite sufficient. There is much wisdom in the famous words attributed to Isaac Asimov, "The most exciting phrase to hear in science, the one that heralds new discoveries, is not Eureka! (I found it!) but rather, 'hmm . . . that's funny . . .'".

After this discussion of the alleged theory-dependence of observation it might be tempting to ask whether the hypothetico-deductive method really begins with theory? The attentive reader may have noticed that the method assumes that there are already observations to generate hypotheses from. When faced with a seemingly inexplicable phenomenon, we said the scientist should propose speculative theories about its cause. Observation of the phenomenon actually precedes the theories. In the example with the prisms, Newton saw that they disperse white light into several colors *before* he thought of recombining the colors. Semmelweis noticed the difference in mortality between the two maternity divisions *before* working out hypotheses. Theory does not necessarily precede observation. More importantly, there are no pure inductivists or falsificationists among working scientists. These are idealized concepts used by philosophers to make it easier to discuss various aspects of research. From a scientist's point of view, the most important thing to keep in mind when comparing the two approaches is that they lead to different types of knowledge. There is nothing wrong with objective observation as a starting point for the advancement of science, but if we want to explain what we see we must – at some point – formulate hypotheses and test them against reality.

2.7 The Scientific Community

It is very difficult to solve complex problems on your own. Luckily, science is a collective process. Even if you do work on your own you always build on foundations laid by others. Our knowledge would grow too slowly if every scientist were to invent the wheel every time they solved a problem. Therefore, science is also an open process. By sharing their results through publications scientists help each other make progress and all researchers within a field benefit from this. This open, collaborative aspect of research is called the scientific community. When starting out in research it is important to get to know it and to start interacting with it.

The scientific community has many subcommunities. The most obvious one consists of the people that you work with every day, such as your closest colleagues and your supervisor. Another fairly obvious subcommunity is the people working within your subdiscipline, whether it is behavioral ecology, laser spectroscopy, or any other research field. These are the people that your research group collaborates with, and the ones you meet at scientific conferences. Disciplines such as physics and chemistry are on an intermediate level, and the top level includes all people that have ever worked with, and ever will work with, scientific research. Some parts of the scientific community are very closely connected, sharing techniques, theories and problem areas with each other. Other parts are more distantly related. Since science both builds knowledge and transfers knowledge between people it can be seen as a web that interconnects all nations and all times. Every node of the web is a scientist, every thread is a transfer of knowledge. When you read scientific papers you can follow those threads by looking at the reference list at the end. Scientists who have made a great contribution to science have more threads connected to their node but each and every one who contributes to science is part of the web. I have never tried it, but I am positively sure that it is possible to follow the threads back in time through the web from any researcher today to the pioneering scientists during the renaissance.

When discussing scientific method, it is interesting to ask if research should be considered to be "unscientific" if it falls outside the inductive or hypothetico-deductive framework. Much research does. Some scientists get research grants and academic prizes for developing new measurement techniques, without ever explaining any natural phenomenon at all. Others use routine techniques to map out some aspect of nature. For instance, physicists who investigate the structure of atoms use established spectroscopic techniques to collect data. They proceed to interpret their spectra by established analysis techniques. The results are published in databases and are accepted as good quality research within their field, despite the fact that not a single bold hypothesis has been formulated or tested. Some philosophers of science maintain that such results become useless after a theoretical paradigm shift, since the information is incorporated in the definitions and concepts of the existing theory. As we have seen, this is not necessarily true. More importantly, we could equally well maintain that science becomes useless without such results. The quantum mechanical theory of the atomic world is obviously a great scientific breakthrough but if it were not used to analyze the structure of atoms in this way we would not be able to use atomic spectra to other scientific ends, such as analyzing the composition of distant stars. This would obviously be an impediment to our ability to develop theories about the evolution of stars. Without the tedious everyday work with established theories, the theories become museum pieces, to be admired but never put to use. Science builds on itself, using established theories to develop new knowledge. New theories are not developed by individuals but by communities. Science is a much more complex, inhomogeneous, and multifaceted activity than inductivists and hypothetico-deductivists wish to let on. And even though not all science is conducted in the same manner, all science is nonetheless interconnected.

If scientists do not adhere strictly to the methods proposed by philosophers, where do they learn scientific method? In many places, a course in the philosophy of natural science is a healthy part of the Ph.D. curriculum, but the major part of the training to become a scientist takes place elsewhere. Scientific method is learnt where science is being made, under supervision by experienced researchers. It involves a wide variety of skills; the craft of operating experimental apparatus, sometimes also of designing and building this apparatus, knowing how to create experimental conditions that make it possible to obtain useful information, the craft of acquiring and interpreting data, and so forth. As a budding researcher you also learn a process of working in parallel with facts and ideas to solve research problems, generating ideas from facts, comparing ideas with facts. You learn a combination of craftmanship and mindset that, in time, will enable you to contribute in peer review processes yourself. In practice, it is these reviews by fellow scientists, and not the philosophy books, that judge what good scientific praxis is.

In this chapter we have attempted to understand what science is. We have seen that this is no easy task. Quite possibly, we are now faced with more questions than we started with. This, on the other hand, is a natural consequence of reflection on any problem. The next chapter is about how it happens that we have science at all. This is by no means as self-evident as it may seem.

2.8 Summary

- The ultimate goal of science is to explain the world. Observations should, therefore, be related to theories to make a lasting contribution to science.
- In the inductive approach, science starts with unbiased observations from which general theories are developed. Such theories cannot be proven in a logical sense. Critics of the approach maintain that observations always are biased, since they rely on theory in some form. The theories obtained by induction have no explanatory power.
- In the hypothetico-deductive approach, science starts with an hypothesis, a tentative theory, which is to be tested against observations. Hypotheses that do not survive the tests are discarded. The others may become established theories with time. Neither these theories can be proved logically, but they have explanatory power. A disadvantage of this method is that it seems counter-intuitive to many scientists, who tend to think of the progress of science as governed by confirmation of theories, rather than falsification of them.
- Observation and experimentation have more central roles in practical science than in the inductive and hypothetico-deductive approaches, and may even exist independently of theories. Theory-independent observations often play an important role in the discovery of new phenomena but they must be related to theories to explain the phenomena.
- Observation is a skill. Our powers of observation can be improved by training and experience. This does not imply that skilled observers are biased. We are not required to give equal weight to all facts to be objective.

- Science is an open, collective process. The participants in the scientific community contribute to and criticize each other's work. Scientific method is complex, since different parts of the scientific community may employ different approaches to research, while all parts are still interconnected.

Further Reading

Chalmers [3] gives a lucid introduction to the major developments in the philosophy of natural science during the twentieth century. It is recommended as a first book on the topic and the interested reader will be guided to some influential books through it. It does not allow much space for observation and experimentation. For those interested in these topics, the second half of Hacking's book [8] provides a coherent treatment of experimental science, including a large number of nice examples.

References

1. Molander, B. (1988) *Vetenskapsfilosofi*, 2nd edn, Thales, Stockholm.
2. Phillips, E.M. and Pugh, D.S. (1994) *How to Get A PhD: A Handbook For Students and Their Supervisors*, 2nd edn, Open University Press, Buckingham.
3. Chalmers, A.F. (1982) *What Is This Thing Called Science?* 2nd edn, Hackett Publishing, Indianapolis.
4. Crease, R.P. (2004) *The Prism and the Pendulum*, Random House, New York.
5. Hempel, C.G. (1966) *Philosophy of Natural Science*, Prentice-Hall, Englewood Cliffs (NJ).
6. Panek, R. (2000) *Seeing and Believing*, Harper Collins, London.
7. Shea, W. (1998) Galileo's Copernicanism: The Science and the Rhetoric, in *The Cambridge Companion to Galileo* (ed. P. Machamer), Cambridge University Press, Cambridge.
8. Hacking, I. (1983) *Representing and Intervening*, Cambridge University Press, New York.
9. Cobb, C. and Goldwhite, H. (1995) *Creations of Fire: Chemistry's Lively Histroy From Alchemy to the Atomic Age*, Plenum Press, New York.
10. Priestley, J. (1775) *Experiments and Observations on Different Kinds of Air*, Thomas Pearson, Birmingham.
11. Dawkins, R. (writer) (2008) The Genius of Charles Darwin [television broadcast]. Channel Four, UK.

3 Science's Childhood

There are indeed two attitudes that might be adopted towards the unknown. One is to accept the pronouncements of people who say they know, on the basis of books, mysteries or other sources of inspiration. The other way is to go out and look for oneself, and this is the way of science and philosophy.

—Bertrand Russell

Today, the fruits of scientific progress surround us to such a degree that we take science and its applications for granted. It may even be difficult to imagine that there was a time when we did not have science. The aim of this chapter is to give a brief account of how the scientific approach came about, as this gives a slightly different perspective on scientific thinking than the previous chapter did. Exploring things from different perspectives makes it easier to understand their essential characteristics. Many of the ideas described will seem odd to our modern eyes and some even irrelevant to modern science. We should remember, though, that they are still important elements in a remarkable change in thinking that has profoundly transformed our lives.

We tend to take science for granted but it is striking how many of the central developments in its history depend on pure coincidences. It takes more than a brilliant mind for a brilliant idea to become successful. That mind has to work under favorable conditions in the right type of society, and it must interact with the right other minds in order to have an impact. To me, it seems like we are indebted to a handful of exceptional people and circumstances for the scientific revolution and the remarkable developments in technology and medicine that followed in their wake. Of course, space is too limited here to provide a satisfactory overview of these people and circumstances and further reading is therefore suggested at the end of the chapter. A timeline is also provided towards the end of the chapter to put the developments into perspective.

3.1 Infancy

When holding a newborn in your arms an avalanche of impressions and thoughts wells forth in your mind. The thought of holding a brand new human being is incredible. The baby seems fragile and helpless, yet the thought of the arduous birth that it has just gone through bears witness to an immense strength. Sometimes, people who have just become parents say that it feels as if their lives are split into two parts when they first see their child. Until that moment it has been difficult to imagine themselves as parents, however much they tried. But when the child lies in their arms, peering at the daylight, it is difficult to imagine that they ever lived without it.

Experiment!: Planning, Implementing and Interpreting, First Edition. Öivind Andersson.
© 2012 John Wiley & Sons, Ltd. Published 2012 by John Wiley & Sons, Ltd.

They say that the great chemist and physicist Michael Faraday was once asked what the use of science was. He replied: what is the use of a newborn child? Of all questions that cross the minds of new parents, this is certainly one of the less common ones. On the one hand, it is true that every human born is useful. Just think of everything that a new pair of hands will be able to accomplish. Still, the question is absurd, since we do not bring children into the world for the sake of usefulness. There is more to life than productivity and, above all, we hope that our children will be happy.

Just like a child, science carries with it a promise for the future. During the relatively short part of our history that we have occupied ourselves with science our lives have changed entirely. Learning to understand the world around us has made us able to fight diseases, tame the energy flows of Nature and use her processes to create things that our ancestors could not even dream of. There are few ideas, if any, that have changed our existence so fundamentally as science. But the first of our forerunners that were devoted to science did not do it for the sake of usefulness. They did it out of curiosity and for the joy of learning to understand the workings of the world. They could not imagine what their curiosity would bring with it in the long run. If we ask scientists working today where they find their inspiration they will give us the same reasons. We expect more from life than just being productive.

As a child changes the lives of its parents, science has fundamentally changed the conditions of humanity. Though the boundary is not sharp we can, in some sense, divide our history into the period before the birth of science and the present epoch. Scientific inquiry has given us the technical progress that we now take for granted and technical development constantly yields new problems to explore scientifically. Developments in technology and medicine go hand in hand with a scientific view of the world. Without this view our living standard would surely not have increased as drastically from the medieval level as it has. In this chapter we are going to ask how it happened that we invented science in the first place.

3.2   Ionian Dawn

Our voyage to the roots of science begins on the shores of the Aegean Sea. Today, the Greek archipelago is mostly known as a destination for tourists. We go there to relax, but many of us also get a strong feeling of being on historic ground by the Mediterranean. The ancient Greeks are associated with wisdom and are said to have laid the foundations of modern civilization, but who were they? Why do we still speak so reverently about them? We know that they were an enterprising, seafaring people that had an intense exchange of goods and culture with other peoples. Those who walked in the ancient docklands of Greece experienced an impressing buzz of foreign languages and new ideas – a cultural draught of rare occurrence.

Maybe it was this diversity that led to a special awakening on the coastlands of Asia Minor a few hundred years before the Common Era (BCE). Here, there were suddenly people who held that everything consisted of tiny building blocks called atoms. A man lived who lived here reasoned that humans must have developed from other animals and thought that life must have occurred spontaneously in the seas. There were people who believed that illnesses were not caused by gods or evil spirits but by natural causes. One of these islanders thought that the earth was a planet orbiting the sun, and not vice versa. Here were people who stated that the world's countries had come about not by intervention by higher powers but by natural, geological processes. Today, we see these ideas as relatively modern but, in reality, they are two and a half millennia old. In the following, which loosely follows the outlines of Farrington [1] and Russell [2], we are going to get to know a few of the persons behind these ideas.

Although the Mesopotamians and Egyptians had developed technology and mathematics to a certain level, what we call the scientific approach first arose in the Greek culture. The older civilizations were agricultural societies with strong central powers. Mathematics and astronomy were used for purely practical purposes, to measure the land and create calendars to facilitate farming. Religion and politics were intimately interwoven and the rulers wanted to be perceived as demi-gods. They had a strong interest in maintaining the prevailing order and religion was an instrument for doing it. The clergy preserved a dogmatic religion. Such societies hardly encourage pioneering thoughts and the questioning of established truths.

The Greek world, by contrast, was not at all as centrally controlled as the older civilizations. The landscape was barren and this made communications difficult. Instead of a central state authority, a large number of autonomous city-states were formed. As people lived from trade there was a strong interest in innovation. The merchants that governed the societies actively supported the technical developments that made business flourish. The cultural exchange also contributed to develop critical thinking. What do you do when you discover that there are two gods who claim to be the ruler of heaven? The Babylonian god Marduk played this role in Mesopotamia, just as Zeus did in Greece. Their differences suggested that they were not the same god with different names. Was one of them just a figment of the human imagination? But if one was, why not both? [3]

Thales counts as the first scientist. That is not just because he acted like an absent-minded scientist: they say that he tripped and fell into a well one evening when he was walking and gazing at the stars. A slave girl is said to have laughed at him for being so concerned with what was in the sky that he did not see what was at his feet. He lived in the sixth century BCE in the town of Miletus on the Aegean coast of what is now Turkey. Besides this anecdote it is his practical and rational attitude towards the world that has given him fame. Both the Egyptian and Babylonian religions imaginatively explained how the world once had arisen from the water. The thought was probably natural, as both countries had literally been conquered from the water. The Babylonians belived that Marduk had bound a rush mat on the water and created the earth on it. Thales also thought that everything had once been water but he left Marduk out of the picture. He held that the world had arisen from the water through a natural process, similar to the silting up of the Nile delta. He describes the world as a disc on the water where we have water on all sides, also above us – where else would the rain come from? [1]

What makes this water hypothesis scientific? The answer lies more in his approach than in his statement. By leaving Marduk out he searches for natural explanations for natural phenomena, and this is what is new and essential. He tells us that things do not happen through the interventions of Zeus or Marduk; it is possible for us humans to understand and explain nature. The water theory may seem bizarre to our modern eyes but is it completely unreasonable? Neither plants nor animals can live without water. Water falls as rain from the skies to the earth and flows via the rivers out to the sea. It can evaporate and become steam, or freeze to ice, so the thought that everything is a circulation of water in different forms seems quite natural. It is not far from the modern view of hydrogen as the origin of all elements. The hydrogen atom is by far the most common one in the universe. When pushed together under unimaginable pressures in the interiors of stars, hydrogen atoms merge into atoms of helium and heavier elements in the nuclear processes that generate the light and heat of the stars. Every element on earth that is heavier than hydrogen – even the atoms in our own bodies – have been spread across the universe by stars that exploded billions of years ago. Our solar system, our earth, and we ourselves are made of this star stuff. Thales' idea of a single principle behind all matter is certainly not far off.

Thales made trips to Egypt at an early stage; here he learned geometry. He elaborated the Egyptian geometry and, among other things, developed a method for determining the distance to a ship at sea. He realized that it was possible to measure things, even though they were out of reach. We also owe him for another essential development of geometry. The Egyptians already knew that a circle is split into two equal parts by its diameter, but Thales could give a general *proof* of it. He realized that geometry could be used for more than solving specific problems, it could also be used to obtain general knowledge [1].

He is also famous for predicting a solar eclipse. It is not likely that he had a functioning theory for solar eclipses, so he probably used Babylonian tables and was lucky enough that the eclipse was visible from Greece. At any rate, the eclipse has been dated to 585 BCE and by that we know when he was active [1].

As we have seen, people were mocking Thales for philosophizing instead of engaging in practical things, but he got even with his mockers. One year his grasp of meteorology told him that the olive harvest would be unusually rich and he hired all the olive presses he could lay hands on. At the time of harvest he let them at his own price, making a round sum of money. It was indeed possible to make money on philosophy, he said, but that was not why he engaged in it [2]. In addition to being a good story, this anecdote tells us something important about the times and the society in which Thales lived. You were supposed to engage in things that were practically useful. A couple of hundred years later the times would have changed to the degree that practical relevance was not even desirable. During Thales' times slavery had not developed enough for practical work to be despised. Scientists, or natural philosophers as they called themselves, worked with practical things. Many of them were sons of sailors, farmers, and weavers. Those who speculated about the laws of nature based their conclusions on everyday practical experience. Craftsmen, engineers and scientists talked to each other and were often the same persons. Among many other things, these Greeks invented the spirit level, the potter's wheel, bronze casting, and central heating [3]. They knew from experience that Nature follows regular patterns that could be discovered and applied to practical purposes. From their perspective it was sensible to seek natural explanations for natural phenomena and take the gods out of the equation.

Before continuing, it may be appropriate to reflect on what this means. If we say that science is born when humans first try to explain the world without referring to deities, does it mean that there is a deep chasm between religion and science? The answer varies between researchers. It is, perhaps, more constructive to reformulate the question: is it possible to be both religious and a successful scientist? The answer is a resolute yes, because there are plenty of examples of such people. If we choose to see science as an approach for solving certain types of problems it is, naturally, possible to use that approach regardless of one's personal philosophy. It is just as possible for a religious person to do research as it is for an atheist to practice yoga. Believing that there is something beyond the observable, some deeper reason behind the regularities in nature, does not necessarily conflict with the wish to investigate and understand these regularities. But if this something really lies beyond the observable, inaccessible to investigation, it also lies beyond the realm of evidence-based knowledge. Whatever view we take to science, it is important to note that gods and miracles are conspicuous by their absence in scientific theories. Not even deeply religious researchers would dream of explaining their data by the intervention of higher powers.

Our next natural philosopher from Miletus, Anaximander, was a friend of Thales. Just like Thales he reflected over practical things and interpreted Nature starting from everyday observations. He did not agree that water was the first principle because he thought that the basic stuff of all matter could not be one of its forms. He chose to call the first principle "the Boundless". He imagined the sun, moon, and stars to be made from fire and that they were visible through holes moving along the vault of heaven. Anaximander is also the first person known to think that humans had evolved from other animals. He comes to this conclusion by an amazingly simple and elegant line of reasoning. Since our children are born so helpless that they are not able to survive on their own, he says, there would be no humans today if the first of us had come into this world alone. We must have evolved gradually from animals with more autonomous offspring. He thought that life had originated in the seas and that we were descended from fish. Some of the fish had adopted to a life on land and developed into other animals. He supported this statement with fossils and the observation that sharks feed their young. Surely, this idea is what makes Anaximander urge us not to eat fish. When Bertrand Russell describes this idea he jokingly adds: "Whether our brethren of the deep cherish equally delicate sentiments towards us is not recorded" [2].

The last Milesian that we are going to discuss is Anaximenes. According to him, air is the basic element from which everything is formed. This may not seem like a completely fresh idea but his theory contains something new and interesting. The keywords are condensation and rarefaction – words that describe physical processes. He thinks that air becomes heavier and changes form by condensation. By rarefaction it becomes lighter. Anaximenes is said to have gotten the idea while watching water evaporate and condense. He held that condensation was associated with cold and rarefaction with heat. If you have taken a course in thermodynamics you know that this is, in fact, quite the opposite of the truth, but Anaximenes does not expect us to just accept his statements. He shows us how we can convince ourselves by a simple experiment. Breathe against the back of your hand with wide open mouth. Doesn't this rarefied air feel warm? If you instead form your lips to a small opening and press condensed air through it, the feeling on your hand is cold. Of course, Anaximenes did not know what lies behind this effect. Do you? [1]

There are several more natural philosophers in the Ionian tradition that deserve mentioning. This chapter is much too short to do them justice, but we should at least mention a couple of their ideas. Democritus, from the Ionian colony Abdera in northern Greece, said that on the smallest scale of the world there is some irreducible degree of graininess. To describe this he invented the word atom, meaning something that cannot be split. He maintained that there is nothing else in this world than these indivisible atoms and empty space. Therefore, when we cut through an apple the knife must move through the void between the atoms, otherwise it could not be split. He also said that when a cone is cut in half, the two section surfaces will have somewhat different areas due to the tapered shape. By a similar line of reasoning, Democritus concluded that it must be possible to calculate the volume of a cone by adding a number of very thin plates with diminishing diameter from the bottom to the top. By this he came close to inventing infinitesimal calculus, a central mathematical tool that was key to the revolution in physical theory in Newton's time [3].

Finally, for those who think that Nicolaus Copernicus invented the heliocentric theory, in which the earth revolves with the other planets around the sun, it may be interesting to

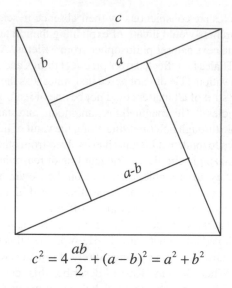

$$c^2 = 4\frac{ab}{2} + (a-b)^2 = a^2 + b^2$$

Figure 3.1
A simple, geometric proof of Pythagoras' theorem. Calculating the area of the large square by adding
the areas of the four right-angled triangles and the small square, it directly follows that $c^2 = a^2 + b^2$.
It is not known if Pythagoras himself could prove the theorem.

know that a marginal note of the name Aristarchus in one of his manuscripts reveals that he
only rediscovered an idea from antiquity. Aristarchus lived in the third century BCE on the
island Samos, near Miletus. Although his original manuscript has been lost, Archimedes
tells us in his book *the Sandreckoner* that Aristarchus advocated the heliocentric view.
Though it cannot be ruled out that the idea is even older, Aristarchus appears to be the first
to give a full and detailed account of this system [2].

3.3 Divine Mathematics

When later philosophers describe the early Greek attempts at a scientific explanation of
the world, they distinguish between two schools of thought. On the one hand there is the
Ionian school that we have discussed. They sometimes call it atheistic, as it relied only on
material causes. The other school is a religious one, arising in greater Greece in the west.
In Plato's account, the Ionian school does not recognize any form of intentional creation
in Nature. It claims that only the human mind creates intentionally and it, too, arises from
material causes. The religious school, by contrast, claims that purpose and thought precede
the properties of matter. Nature is ruled by reason [2].

   This religious tradition begins with Pythagoras, who is known to most school children
through his famous theorem. Figure 3.1 gives a simple geometric proof of it. He was
originally from the island Samos, just a stone's throw off the coast from Miletus, but he
moved to Croton in southern Italy where he founded a religious community. They practiced
asceticism and taught that it was, for some reason, sinful to eat beans. They also spent time
contemplating mathematics, where they saw the key to understanding the world.

Scientifically, the Pythagoreans were highly successful in number theory and it is really the mathematicians that are their true inheritors today. But the Pythagorean idea that the world can be understood using mathematics has had a profound influence on the development of natural science – a fact that can be verified by anyone who has looked in a physics textbook. Still, their variety of science constituted a break with the Ionian tradition. It based itself more on abstract ideas than on information provided by the senses. Philosophy had been practical but with the Pythagoreans it became a detached contemplation of the world [2].

The Pythagorean cosmology seems odd to modern eyes. For them, mathematics was more than a means to describe the world. They literally believed that the world consisted of numbers. Their points had size and could be added to lines. Lines had width and could be added to surfaces, and so forth. To them, number theory was physics. But mathematics also had a moral quality. For example, since the celestial bodies were thought to be divine the Pythagoreans declared that they *had* to be perfect spheres, moving in perfect circles. Nothing else would be suitable in the heavens. This thought would have a tremendous impact on later thinkers and obstruct real advances in astronomy for two millennia [1].

The Pythagoreans discovered the simple numerical relationships that are known as musical intervals. When the string of an instrument was shortened to half its length its sound was raised by an octave. Other fractions of the length produced other intervals. The fact that they could describe the sounds of a string mathematically is probably the origin of their belief that there is a mathematical structure in the universe [2]. They were also fascinated by regular solids, which are bodies where all sides consist of identical regular polygons. There are only five such solids and they are shown in Figure 3.2. For some reason, the Pythagoreans thought that knowledge about one of them, the dodecahedron, was too dangerous to be revealed to people outside their community. The four other regular bodies were connected with the four elements that were believed to make up this world. They thought that the fifth must be connected with a fifth element, existing only in the heavens [3].

A crisis occurred within the order when it became clear that the diagonal of a square could not be expressed as a ratio of two numbers. According to the Pythagorean worldview a line consisted of a finite number of points and the finding showed that this idea was incorrect. It is especially ironic that the discovery was made using the Pythagorean theorem [3]. Their immediate reaction was to forbid spreading the discovery. The story goes that one of the disciples was drowned at sea for divulging the secret [2].

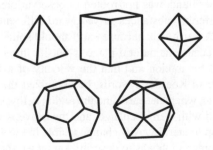

Figure 3.2
The five Pythagorean solids.

The mathematical worldview would also give rise to the theory of ideas. This theory would have a great influence on Plato, who will be introduced below. When talking about a triangle, for instance, they did not mean a triangle drawn on paper. Such triangles are always imperfect due to practical reasons. Lines can never be drawn quite straight with pen and paper, and so on. For this reason, the laws of geometry only fully apply to a perfect triangle, or the *idea* of a triangle, which exists only in the mind. In other words, they thought that the world of ideas contained pure, perfect things, whereas the real world could only produce second-rate representations of them. This line of thinking would later develop into the notion that only the mind could access the perfect, real, and eternal. The sensible was apparent, defective, and transient [2].

Parts of the Pythagorean heritage were to inspire the development of modern science and still have relevance today. The Pythagorean's were the first to use the word cosmos for the world, which means order. The development of physics owes much to their idea of a mathematical structure in the cosmos. But in other ways their departure from the Ionian tradition was to obstruct the sprouting seeds of science. Observation and experience became subordinate to abstract thinking. They based their science on gentlemanly speculation and were reluctant to experiment. From one perspective, turning your attention away from the material world probably seemed right in a society where the abhorrence for physical labor increased at the same rate as the slavery. But, when trying to understand the world, it is not a fruitful approach to ignore the facts and work solely with preconceived ideas. Furthermore, keeping inconvenient truths secret from the rest of the scientific community does not promote the growth of knowledge.

The religious tradition of the Pythagoreans was continued by two of the most famous philosophers of all time, Plato and his student Aristotle. Their ideas, in turn, have had a tremendous impact on western thinking. One important reason for this is that a large portion of their works has survived. We only know the work of earlier philosophers by fragments of manuscripts and references by other authors, but in Plato's case we have the complete works [4]. Another important reason may be that Aristotle was the private teacher of Alexander the Great, who was to spread Greek culture widely over the world. Although they are important as philosophers, they are often considered to be less interesting as scientists. This is especially true of Plato, who maintained that contemplation of the abstract is a safer way to knowledge than actual observation. He believed that the physical world was riddled with imperfections, whereas the world of ideas was perfect.

Some parts of Aristotle's work were influenced by this view but in other parts he clearly recognized the value of observation. He made great progress in biology, mainly by describing and categorizing information obtained by observational studies. His physics, on the other hand, was not rooted in observations of the real world. In summary he thought that different substances strive toward different layers in the universe to find their "natural places". Earth falls because earth should somehow be at the center of the universe and fire rises because its natural place is outside the other elements. He also taught that force is needed for motion and that the velocity of a falling object is directly proportional to its weight. In Koestler's words, he organized the Platonic theory of ideas into a two-storey universe, where the earth and the region below the moon were seen as corrupt and defective, plagued with eternal change, and the heavens were of another substance, immutable from creation to eternity. Any change visible in the heavens could thereby only be apparent. The astronomer's job was to describe, on paper, the apparent motions of the planets and not to offer real explanations. The job was to "save the appearances", for what was there to explain when nothing changed? Ptolemy elaborated the Aristotelian cosmology by adding deferents and epicycles to the heavens to force the planets to move in circular, uniform, "perfect" orbits. This cosmology would maintain its stranglehold on the progress of astronomy through the middle ages and, scientifically, this period would be less than spectacular [4]. Therefore, we will now quickly leap nearly two millennia forward.

3.4 Adolescence – Revolution!

Looking through the history schoolbook of my upper teens, most of it covers the doings of kings, generals and other people of power. The occasional digressions from this theme mostly deal with developments in the arts, sometimes in technology and, more rarely, in science. Out of about four hundred pages, only two are bestowed on what is sometimes called the Great Scientific Revolution. In my book, this turning point in our history goes under the more modest title "Scientific Advances". I recently found that I, as a pupil, had put these two pages within parentheses, along with a penciled note of my history teacher's recommendation: "never mind this".

Today, the remnants of the political intrigues in my history book mainly consist of new names and borders of nations. Although history's perpetual wars had the direst possible consequences for the people who had to endure them, the lasting effects on our present lives are limited. Regardless of the outcome of the Thirty Years' War, for example, people living in catholic and protestant parts of Europe now enjoy similar health and living standards. Considering the radical effects that the scientific revolution has had, and still has, on our everyday lives, the outcomes of the political schemes of days gone by are dwarfed in comparison to the achievements of the scientists of the renaissance. We are greatly indebted to them and I wonder why many history teachers miss this point. I venture to think that it is only a matter of disinterest in science. For those who, like I, never learned about the Great Scientific Revolution, here is a quick recapitulation of one of the most wide-ranging developments ever to occur in human thinking. It loosely follows Koestler's [4] account. At the end of this chapter we will also discuss how it has changed our lives, but let's start at the beginning.

During the Middle Ages, Aristotelian physics and astronomy had become part of the official doctrine of the church, which had grown into a significant political force. As its power was believed to emanate from the highest possible quarters it was surely advisable not to question its authority. As a result, the scientific authority of Aristotle remained unchallenged. But throughout this period the Arabs had preserved the heritage of other Greek thinkers. Eventually, their long forgotten texts began slowly finding their way back to Europe. It is in the wake of this rediscovery of ancient ideas that the Great Scientific Revolution began [4]. The story begins with a Polish canon called Nicolas Koppernigk, better known under his latinized name Nicolaus Copernicus, who left his homeland to study in Italy around the year 1500. Here, he came into contact with both Pythagorean writings and ancient astronomical ideas, most notably the heliocentric theory of Aristarchus of Samos. These ideas stayed with him throughout his entire life and he elaborated them into an astronomical system in his book *On the Revolutions of the Heavenly Spheres*. In this, his only scientific work, he puts the earth adrift with the other planets around the sun, which he puts at the center of the universe. The Copernican system is shown in Figure 3.3. Knowing from the start that these were deeply inappropriate ideas he delayed the publication some thirty years. The first copy of the book arrived from the printer only hours before his death

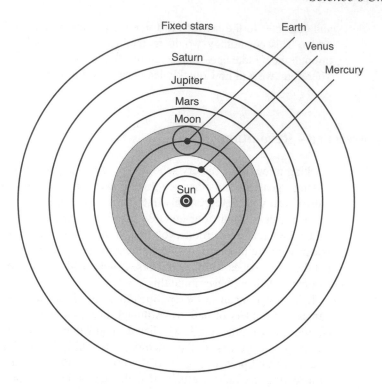

Figure 3.3
A schematic representation of the Copernican system. It has the sun at the center and places the orbits of the then known planets around it, in the correct order.

in 1543. His suspicions were confirmed: the book was to remain on the church's list of prohibited books until 1758.

There is a popular notion that Copernicus instantly gets things right, that his book takes astronomy out of the medieval darkness and puts it in the modern era in one single leap. But his theory is not without problems. It is still dictated by ancient, preconceived ideas. The planets are still moving uniformly in circular orbits, the only thinkable motion of a heavenly body. Throughout the Middle Ages it had become standard practice to add epicylces to the orbits whenever the Ptolemaic system did not fit the astronomical observations. Copernicus also uses epicycles to iron out the wrinkles that arose from the fundamental errors that were still present in his system. Still, his system was fundamentally more correct and that made it simpler than the Ptolemaic one. Less patching up was needed to cover up the errors [2]. Copernicus' main contribution to the development of science was to revive the great idea of Aristarchus of Samos and to make it accessible by publishing it in his book. He did not make many astronomical observations of his own but relied almost exclusively on Ptolemy's data. Theoretically he was reusing most of the ancient ideas [4].

The person who made the crucial contribution to the revolution of science was a German school teacher named Johannes Kepler. He began thinking about Copernicus' universe while still a theology student during the 1590s. For him, the central sun was a symbol of God, the source of heat, light, and the force that drives the planets round their orbits. He also favored the Copernican system because it was geometrically simpler than the Ptolemaic one. Kepler too was heavily influenced by Pythagorean thinking and was to build his own

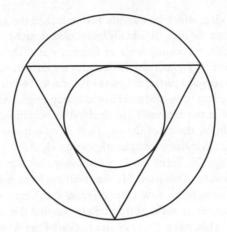

Figure 3.4
When Kepler drew this figure on the blackboard it struck him that the ratio of the two circles was the same as that of the orbits of Saturn and Jupiter.

model of the universe around the five Pythagorean solids. The influence is also evident in his later idea of a universe governed by musical harmonies. While being a teacher in the Protestant School in Graz, Austria, he was riddled by questions about the universe that seem completely meaningless to us. Why were there only six planets? (In his days no planets were known beyond the orbit of Saturn). Why were the distances between the planets and the sun what they were? Why did they move at their particular velocities? He made endless attempts to solve these problems.

Eventually, while giving a geometry class, he had an epiphany. He had drawn an equilateral triangle and two circles on the blackboard, as illustrated in Figure 3.4. One circle was inscribed in the triangle, the other was circumscribed around it. In a flash it occurred to him that the ratio of the two circles was the same as that of the orbits of Saturn and Jupiter. There seemed to be a geometrical structure to the cosmos! Though he failed in applying this idea to the other planets he felt that he was on to something. As two-dimensional polygons could not be matched to the planetary orbits he shifted his interest to three-dimensional shapes. The reason that there are only six planets, he thought, must be that the five Pythagorean solids determine the intervals between the orbits. And, surprisingly, he found that the solids fitted the orbits. Or at least they seemed to fit, more or less [4].

At the age of 25. Kepler published his innovative model of the cosmos in *Mysterium Cosmographicum*. In the first part of the book he works out a "principal proof" that the planetary spheres are framed in the five perfect solids. After this argument, based on preconceived ideas and divine inspiration, something remarkable happens in the text. The five solids, he says, is only an hypothesis that needs to be backed up by observational data. "If these do not confirm the thesis, then all our previous efforts have doubtless been in vain" [4]. In one leap he jumps from vague metaphysical speculation to the position that the truth is to be determined by observation of the real world: from the Pythagorean track back to the spirit of the Ionians. Like the Sleeping Beauty, the Ionian approach to science wakes up after being dormant for almost two millennia. This leap put Kepler in an intellectual dilemma, as he was soon forced to conclude that the data and the model did not match. Was there a problem with the model, or were Copernicus' data to blame? Enamored with his great idea, which he thought had unlocked the grand mystery of the cosmos, he preferred to blame the data. To obtain better observations, Kepler set his hopes on Tycho Brahe.

Tycho, a contemporary of Kepler, was a Danish a nobleman who had originally been destined for aristocracy. Luckily for us, an astronomical event intervened. In his early teens he witnessed a partial eclipse of the sun that would change his life entirely. The fact that the eclipse had been predicted made a deep impression on him. How was it possible to foresee events in the heavens? He decided to become an astronomer and his genius manifested itself in the design of the most advanced astronomical instruments that the world had seen before the invention of the telescope.

At age 26, Tycho made a discovery that was to establish his reputation as the leading astronomer of his time. He was walking from his uncle's alchemy laboratory one evening when he suddenly saw a star brighter than Venus where no star had been before. His unique instruments allowed him to determine that the new star was much further away than the moon. This was a direct contradiction of the Aristotelian doctrine that changes could only occur in the region below the moon and that the sphere of the fixed stars was invariable. Tycho published the discovery in his first book *De Nova Stella*, which has given us the word "nova". It is still used to describe a class of exploding stars although, technically, Tycho's star was a supernova.

Tycho was an eccentric character. For example, he kept a tame elk at his estate. In place of his nose he wore a golden prosthetic, as he had lost his own nose in a student duel over who was the better mathematician. In his youth, he studied at many universities all around central Europe and finally considered settling in Basel. However, word of his fame reached King Frederick II, who decided to entice him to return to Denmark. In 1576, the king sent a royal order for Tycho to come and see him at once and made him an offer that he could not refuse: the entire island of Hven in the Øresund strait, along with means to build a castle and observatory there. He was also offered a regular income that was among the highest in Denmark. He accepted and built the Uraniborg Castle on the island, the first building ever built with astronomical observation as its primary design criterion. It was a grandiose edifice and the only thing missing in it was the tame elk. While being dispatched to the island the animal had spent a night in a mainland castle, where it had wandered up the stairs and had too much beer to drink. As a sad result it had stumbled in the stairs, broken a leg and died [4].

As time went by, Tycho's manners made him ever more unpopular on Hven. His relationship with the next king, Christian IV, did not develop well and his income was decreased. Being used to a cosmopolitan lifestyle, Tycho also became restless on his remote island. As the circumstances changed his wanderlust took over and he eventually left for Prague, where he became Imperial Mathematicus to the Emperor Rudolph II [4].

It was at this time that Kepler set out to meet him. Due to religious persecution of protestants in Graz, foreboding the infamous Thirty Year's War, Kepler was forced to leave Austria. He headed for Prague to see Tycho, only to find that he would have to struggle more with Tycho's eccentric character than with the data that he had longed to get to work with. Instead of giving Kepler full access to the data, Tycho disclosed them to him piecemeal. He was not inclined to just give his life's work to a much younger scientific competitor. But some time later he was taken seriously ill and, on his deathbed, he bequeathed his observations to Kepler [3].

Eventually, Kepler was forced to realize that his theory of the five Pythagorean solids was no more supported by Tycho's data than Copernicus'. He understood that a model based on circular orbits could not explain the planetary motions. By shifting from circular orbits to elliptic ones he got a much better match with the data. In time, Tycho's observations let Kepler formulate his three laws of planetary motion. They state that the planets move in elliptical orbits with the sun in one focus and that they move faster when they are closer to the sun. The last law establishes a relationship between the length of a planet's year and its mean distance from the sun. In these laws the Pythagorean and Ionian traditions merge. Observed facts finally take precedence over speculation and preconceived ideas. On the other hand, the mathematical worldview of the Pythagoreans is justified. The numerical structure of the observations provided the key to understanding the world [3]. With his three laws, Kepler takes an intellectual leap out of the medieval deadlock that had brought developments in physics and astronomy to a standstill. He set us on the path of using exact observations to develop exact scientific theories. Still, his laws are purely descriptive. Though exact, they do not *explain* the planetary motions. Kepler speculated that the planets were propelled in their orbits by invisible spokes emanating from the rotating sun. This is a step forward from the older idea that angels pushed them round their tracks, but not a big one.

It is interesting to note that this crucial development in scientific thinking hinges on a combination of unlikely historical coincidences. Kepler was not well off and had Tycho remained in Denmark he could not have afforded to visit him. He would probably not have gone to Prague either, if he had not been forced from Graz by religious persecution. Apparently, positive things may sometimes come from such afflictions. Furthermore, these coincidences occurred in the nick of time, since Tycho had only 18 months to live when Kepler arrived in Prague. After Tycho's death, Kepler succeeded him as Imperial Mathematicus [4].

Kepler's celestial mechanics described the motions in the heavens. Meanwhile, in Italy, Galileo developed the terrestrial mechanics, describing motions of objects here on earth. This is an interesting story in itself, as Galileo was a pioneer in elaborating the experimental method, without which modern science would be a hopelessly watered down version of itself. Galileo's approach to research is discussed in the next chapter, but before that we will take a look at the final step in the completion of the scientific revolution. This step was taken by Isaac Newton and consisted in merging the celestial and terrestial mechanics into a unified theory. He showed that he same laws that governed the motions of the planets governed the motions here on earth. This would constitute the end of Aristotle and the beginning of modern science.

The development goes back to 1666 when the great plague forced the closure of Cambridge University. The young Newton spent time at home in Woolsthorpe and, apparently, had time to think about a great deal there. The year is often called Newton's year of miracles, as he developed four seminal theories while waiting for the university to reopen. In Chapter 2 we briefly touched upon his theory of optics. The idea of universal gravity was another one. It is probably difficult for us to understand how courageous and grand that idea actually was. It states that every object in the universe, including you, attracts every other object with a force that decreases inversely with the square of the distance between them. This is indeed one of the most magnificent generalizations ever made by a human mind, linking everything to everything else in this cosmos in an unimaginably vast web of invisible interactions. Yet another of his seminal discoveries was differential calculus, the indispensable mathematical tool for attacking this and many other problems that were to be solved in physics over the years to come [4]. As we have mentioned, Democritus had been stumbling close to discovering this method two thousand years before Newton, but now the time seemed to be ripe for it.

Figure 3.5
This apple tree is propagated from a tree that grew in the garden of Woolsthorpe Manor, from which they say the apple fell that inspired Newton's theory of gravity. The original tree is said to have died about 1815–1820. The photograph was taken in the Cambridge University Botanic Garden. © Öivind Andersson.

The idea of universal gravity was sparked when Newton saw an apple fall from a tree outside Woolsthorpe manor. Figure 3.5 shows a tree that is said to be propagated from this very tree. Tradition says that he saw the moon in the sky at the same time, which made him wonder if the force that pulled the apple towards the earth also held the moon in its orbit. He knew about Galileo's work on projectiles and equated the Keplerian orbit of the moon with the Galilean orbit of a projectile, constantly falling towards the earth but failing to reach it due to its high forward speed. If a projectile was fired from a high mountain, he reasoned, it would be deflected from its forward path by the earth's attraction, following Galileo's parabolic trajectory. The higher the initial velocity of the projectile, the farther from the mountain it would land. As shown in Figure 3.6, exceeding a certain critical velocity it would return to the point from which it was fired. He imagined the projectile falling towards the earth at the same rate as the earth's surface curved away from it. Retaining its initial velocity it would thereby go on orbiting the earth just as the planets orbit the sun. In Koestler's words, Newton, by thought-experiment, created an artificial satellite nearly three hundred years before technology was able to implement it [4].

In addition to the theory of universal gravity, Newton developed a general theory of mechanics that described the motions of all objects, in the heavens as on earth. The whole theory was not published until 1687 in his most important work *Mathematical Principles of Natural Philosophy*, or *Principia* for short. The production of this work seems to have been sparked by another of history's peculiar coincidences. Edmond Halley, who is best known for calculating the orbit of the comet named after him, had engaged in a scientific wager that led him to Newton [5]. The wager concerned the explanation for the elliptic orbits of

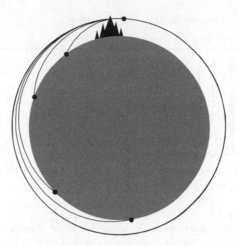

Figure 3.6
Newton's thought experiment. He creates an artificial satellite by firing a projectile with sufficient speed from a high mountain.

the planets. The question so absorbed him that he paid Newton an unannounced visit in Cambridge in 1684. He asked what Newton thought the planetary orbits would be like if they were attracted to the sun by a force following the inverse square law. Newton immediately answered that they would be elliptic. Amazed at the answer, Halley was eager to find out how he could know this. "Why", said Newton, "I have calculated it" [5]. Unfortunately, he could not find his notes there and then, but he soon sent Halley a manuscript where he mathematically derived Kepler's three laws from the inverse square law of gravity. After an extended period of concentrated work, *Principia* emerged. In it, Newton set out his three famous laws of motion and derived from them the new mechanics that applied to every object in the universe. From his law of gravity it directly followed that no angels or spokes were needed to pull the planets around their orbits. All that was needed was the inverse square law.

There it was. The die was cast. Physics had become a mature, quantitative science, able to explain natural phenomena. It was a model for other sciences to follow. This was the beginning of the remarkable scientific development that still continues in our day. To put it into perspective, a timeline is provided in Figure 3.7.

3.5 The Children of the Revolution

Let us return to the question of how this so-called scientific revolution affects our present lives. It is easy to get the impression that what we have discussed mainly is about planets. Their effects on our lives may seem limited, at least to those who do not believe in astrology. But it is important to think beyond the planets because this development is not mainly about them – it is about a revolution in human thinking. This chain of developments is the starting point of modern science and the scientific revolution is, thereby, still very much ongoing. When the Newtonian synthesis had shown what the scientific approach could accomplish, other branches of physics developed in leaps and bounds and other sciences were soon fast

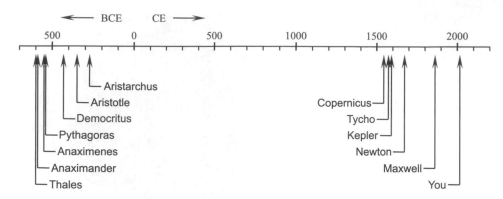

Figure 3.7
Timeline for the characters introduced in the chapter.

on the heels of physics. This chain of scientific discoveries has sparked developments in technology and medicine that would seem like magic to the people of the Middle Ages.

With a little imagination, anyone who has experienced a power failure can begin to picture how the scientific revolution has transformed our lives. Firstly the light goes out, then it becomes cold. After some time we find that it is difficult to cook and that our food spoils when the refrigerator warms up. I do not mean to imply that people starved to death before they had electricity, but due to electricity and other technical achievements we do not have to spend most of our days chopping wood, collecting water, manually washing our clothes, and growing our own food. Since the renaissance, humanity has ventured into a broad program of curiosity-driven research that has paid off in a wealth of convenient applications. For example, our understanding of how infections caused by bacteria can be treated using antibiotics is based on scientific research. Those who live in industrialized parts of the world do not have to worry about many diseases that used to harvest large numbers of lives in the past. Our fast and efficient transports on land and in the air is largely due to the science of thermodynamics, which co-evolved with the development of the steam engine. Our ability to communicate instantly over vast distances by mobile telephones, radio, and televison springs directly from basic research in theoretical physics during the late nineteenth century. Computers and electronics are built from semiconductors, a product of the counter-intuitive theory of quantum physics. Many people connect computers with leisure time activities but let us not forget the phenomenal boost they have given to many areas of research. For instance, biomedical researchers often have to handle vast amounts of information requiring complex calculations that are almost impossible to carry out manually. Large-scale computer simulations are nowadays important research tools across many scientific disciplines. It is also largely thanks to computer control that modern cars emit a mere thousandth of the harmful substances that cars in the 1970s spewed out, to name one example of a practical application. Modern households teem with things that build on discoveries that have followed in the wake of the scientific revolution: washing machines and the detergents used in them, refrigerators, light bulbs and fluorescent lights, microwave ovens, television sets, computers, even lasers in some electronic devices, as well as electricity itself, which powers most of these things.

You may object that these things are technical inventions, not scientific discoveries. Would they not have been invented sooner or later anyway? The answer is that it is difficult to separate developments in technology from those in science, since they often progress in a symbiotic relationship. As I have tried to point out above, innovations that build on scientific discoveries can hardly be accomplished before the basic science that makes them possible is in place. Let us consider an example from Sagan [6] to illustrate this.

Imagine that you are Queen Victoria and, in the year 1860, you have a visionary idea about a machine that will carry your voice and moving pictures of you into every home in the kingdom. This should be done through the air, without wires. You would probably allocate a generous budget to the project and involve the leading scientists and engineers of the Empire in it. Even if such a project probably would produce some interesting ideas and spin-offs, it would almost certainly fail. Although the telegraph existed in 1860, the underlying science that makes radio and television possible had not been developed.

This piece of science would come from a completely unexpected direction. Light had been a mystery since antiquity but James Clerk Maxwell, a Scottish theoretical physicist, discovered that electricity and magnetism unite with each other to make light. The now conventional understanding of the electromagnetic spectrum, ranging from gamma rays via visible light to radio waves, is due to him. So is radio and television, but it is important to note that he did not invent these things – he was only interested in understanding how electricity makes magnetism and vice versa. He made his discovery by intuitive reasoning about the symmetry of the equations that describe electric and magnetic fields, which was partly based on an esthetic judgment. The details are beyond our scope here and, to put it extremely briefly, his speculation was rather reckless and not based on data at all. The implications of his idea were only to be confirmed afterwards in a series of important experiments but, in Sagan's words, this idea "has done more to shape our civilization than any ten recent presidents and prime ministers". The important point is that no matter how much money and time the Queen had provided for her engineers in 1860, they would not have gotten close to inventing television. We know this in retrospect because the scientific foundation for it did not exist [6].

But if scientific discoveries are so important for technical advances, would not the discoveries have been made sooner or later, even without the scientific revolution as a starter motor? This is a tempting and common thought. Since we are surrounded by so much science and technology, we see them as natural parts of our society. But remember that most of the elements needed for the scientific revolution were present already during antiquity. Copernicus had learned from Aristarchus. Kepler had learned from Pythagoras. The technology for building Tycho's instruments was not an insurmountable obstacle. Imagine a world where the idea of the elliptic orbits had been thought of already during antiquity. Picture that Democritus had developed the infinitesimal calculus and made the Newtonian synthesis in the fourth century BCE. It is probable that our civilization would have developed differently in that scenario. Would man have set foot on the moon in the first century BCE? Only three hundred years separate Newton from the United States' Apollo space program, after all. It is, of course, impossible to know if that would have happened. But imagine the opposite scenario instead, that Kepler had never seen Tycho's data, so that the raw material for Newton's synthesis had never come about. Would our cosmology still be built around the Pythagorean solids? Would most of us be out working in the fields instead of reading this book? It is quite possible that we would have had to wait another two millennia for the revolution.

There are no answers to these questions. Had the scientific revolution been born from other ideas in another time, there is no way of knowing what it would have grown up to be.

In that respect, the birth of science is like the birth of a child. The newborn turns a new page in the history book and a new story begins. We do not know anything about what the new life will bring. We only know that a new human being has been born, and that the world will never be the same again.

3.6 Summary

- The scientific approach arose in the Greek culture in the sixth century BCE. The Ionian natural philosophers then made attempts to find natural explanations for natural phenomena, without resorting to higher powers. Their science was based on practical experience and everyday observations.
- According to the Pythagoreans, the key to understanding the world was mathematics. Through this idea they have had a profound impact on the development of physics, though their philosophy based itself more on abstract ideas than on information provided by the senses.
- Aristotle's cosmology was inspired by Pythagorean thinking. In his universe the earth belonged to the region below the moon, which was riddled with imperfection and change. The region outside the moon was constant and perfect. For this reason, the planets must move at uniform speed in "perfect", circular orbits. This influential idea was to become an obstacle to advances in astronomy throughout the Middle Ages.
- Copernicus laid the foundation of the scientific revolution during the renaissance by reviving the heliocentric theory of Aristarchus of Samos.
- Kepler united the Ionian and Pythagorean traditions: he developed a mathematical model of the universe, but it was based on exact observations of the real world. His breakthrough was embodied in three quantitative, mathematical laws for planetary motion. They were accurate but had little explanatory power.
- The scientific revolution was completed with the Newtonian synthesis, which unified Galileo's terrestrial mechanics with Kepler's celestial mechanics. Newton also invented the mathematical method that was crucial for the development of his theory. The synthesis resulted in a general theory of mechanics, applying to all objects in the universe. It explained the origin of Kepler's elliptic orbits and showed that no force is needed to pull the planets around the sun.
- The Newtonian synthesis is the foundation on which the coming developments in physics were built. Newton's breakthrough also served as a model for other sciences.
- Since the renaissance, humanity has ventured into a broad program of curiosity-driven research that has paid off in a more advanced understanding of the world we live in, as well as a wealth of convenient applications.

Further Reading

Sagan's *Cosmos* [3] remains an inspiring introduction to the history of science, as well as to general astronomy. Farrington [1] provides a more detailed treatment of the Greeks, whereas Bryson gives a surprisingly extensive and entertaining overview of the history of modern natural science in his *A Short History of Nearly Everything* [7].

References

1. Farrington, B. (1961) *Greek Science*, Penguin Books, Harmondsworth.
2. Russell, B. (1959) *Wisdom of the West*, Rathbone Books, London.
3. Sagan, C. (1980) *Cosmos*, Random House, New York.
4. Koestler, A. (1964) *The Sleepwalkers: A History of Man's Changing Vision of the Universe*, Penguin Books, London.
5. Sagan, C. and Druyan, A. (1985) *Comet*, Random House, New York.
6. Sagan, C. (1996) *The Demon-Haunted World: Science as a Candle in the Dark*, Random House, New York.
7. Bryson, B. (2004) *A Short History of Nearly Everything*, Transworld Publishers, London.

4 Science Inclined to Experiment

An experiment is a threshold event: It may make use of ordinary and uncomplicated things, but these serve as the bridge to a domain of meaning and significance.

—Robert P. Crease

The cry "I could have thought of that" is a very popular and misleading one, for the fact is that they didn't, and a very significant and revealing fact it is too.

—Douglas Adams

If we want to learn something about a particular phenomenon there are, in principle, two ways to observe it. We may be passive observers, hoping that something interesting will take place while we happen to look. The other way is to deliberately interfere with the world to create a condition that we are interested in. The passive observer is washed over with large amounts of information. Much of it is likely to be irrelevant and it may be difficult to see clear patterns. By creating conditions relevant to a specific question, however, the observer will mostly obtain relevant information without having to rely on luck. This is why experimentation is a more efficient way to obtain information about the world than passive observation.

If you are trying to understand how the planets move around the sun it is, of course, difficult to interact with what you are studying. For this reason, most of the developments discussed in the last chapter were based on passive observation of the heavens. We are still missing a crucial ingredient of modern science: an elaborate experimental method. Galileo Galilei contributed greatly to this development. As he studied how objects moved here on earth, he was able to interact with them and actively seek out the information that was of immediate interest to the questions he wanted answered. He is held to be the first person to obtain a precise mathematical law from an experiment.

The previous two chapters have described scientific thinking in philosophical terms and through its historical development. The present chapter adds another perspective to the list. It introduces the experimental method, which is the main topic of this book, and we will use one of Galileo's most important experiments as a vehicle for this introduction. It shows how a few everyday objects may be used to unveil a fundamental pattern in the workings of the world. Some modern controversies about the experiment will illustrate characteristics of the experimental method and answer a few central questions. What was the aim of the experiment? Was it really possible for Galileo to conduct this experiment? And, most importantly, what *is* an experiment in the first place?

Experiment!: Planning, Implementing and Interpreting, First Edition. Öivind Andersson.
© 2012 John Wiley & Sons, Ltd. Published 2012 by John Wiley & Sons, Ltd.

4.1   Galileo's Important Experiment

Galileo was born in Pisa in 1564, where he grew up in a musical family. His father, Vincenzio, was an esteemed lutist who also made controversial experiments with intonations, intervals, and tuning that challenged ancient authorities [1]. His brother became a professional musician and Galileo played the lute, just as his father. He entered the University of Pisa to study medicine, but soon switched to mathematics. Vincenzio's zest for experimentation seems to have rubbed off on Galileo, because when he got a position at the university in 1589, he began to study the motion of falling bodies.

According to Aristotle, bodies fell at a constant speed that was proportional to their weight. Since free fall typically occurs very fast it is difficult to form an idea of it from direct observation. To be able to study it at all, Aristotle had evidently slowed objects down by letting them fall in water [1]. He had concluded that, in the absence of a resisting medium such as water or air, the speed of fall would be infinite. In Aristotle's view of the world, things fell because they sought their natural place at the center of the universe. He thereby made a difference between the "natural", downward motion of falling objects and "violent", upward motion that he considered contrary to nature.

Galileo showed by thought experiment that Aristotle was wrong in thinking that the speed of fall was proportional to the weight of the falling object. If we suppose that a heavier stone falls faster than a light one, he reasoned, tying the stones together would result in a paradox. The lighter one would then retard the fall of the heavier stone, and the heavier stone would speed up the fall of the lighter one. Despite the fact that they are heavier when tied together they would fall slower than the large stone does on its own [2].

According to tradition, he also demonstrated in a legendary experiment at the Leaning Tower of Pisa that objects of different weight fall at the same speed. The only known account of this experiment arose shortly before Galileo's death and historians of science often doubt that it ever took place. On the other hand, the experiment is so easily performed that it is difficult to imagine that Galileo would have settled with just the thought experiment. Whether he conducted it at the Leaning Tower or not, he probably tried it in one variant or other. In any case, the experiment was not unique to Galileo. Others had performed it before him and shown that Aristotle's ideas about free fall did not hold water. Even though flaws were known in Aristotle's physics, they were generally not considered serious. Galileo's contribution was to demonstrate that the errors were not superficial. Aristotle's system could not be easily fixed – it had to be replaced.

So, Galileo knew that objects of different weight fell through a given distance in the same time. The question was how they acquired their speed. Did it occur instantly at the point of release? Or did they get a series of small and rapid spurts of uniform speed, which had been the common picture during the Middle Ages? The concept of uniform acceleration was not yet known. Galileo realized that the speed acquired by an object in free fall increased with the distance it fell. A pile driver, driving a stake into the ground, doubled its effect if it was dropped from a doubled height. He observed that the weight of the driver stayed the same, so the effect must be due to a change in its speed. In his last book, *Dialogues Concerning Two New Sciences*, he claimed that the falling body acquired its speed through continuous, uniform acceleration. This meant that the object acquired equal increments of speed in equal intervals of time. Since any time interval may be divided into an infinite number of instants, he said, the acceleration must be continuous [2]. In the same book he describes an ingenious experiment from which he obtains a precise mathematical law, describing the motion in free fall. The law relates distances fallen from rest to the times required to cover those distances. Two distances fallen by an object in free fall, s_1 and s_2, are to each other as the squares of the times required to cover those distances, t_1 and t_2, or:

$$\frac{s_1}{s_2} = \frac{t_1^2}{t_2^2}. \qquad (4.1)$$

We would say that the distance is proportional to the square of the time.

The impact of the law was tremendous. Before it, any event in Nature had to be ascertained by observation. Now, an aspect of the world's behavior could be computed with a pencil and paper. It allowed scientists to predict future events and retrodict past events.

Galileo was convinced that slowing a falling object down with water, as Aristotle had done, would disturb the motion more than it would illuminate its nature. Instead of impeding the fall of an object in a resisting medium he extended the time of fall by letting a ball roll down an inclined plane. The greater the angle of inclination, he reasoned, the closer the motion would be to free fall. Decreasing the angle of inclination, the fall would take place over longer times, making it possible to study. By varying the inclination he hoped to find the general characteristics of the motion.

He prepared a wooden board for the experiment by cutting a channel along its edge. It was straight, smooth, and polished. The channel was lined with parchment, which was also as smooth and polished as possible. "After placing the board in a sloping position", he said, "a hard, smooth, and very round bronze ball was rolled along the channel". He noted the time required to make the descent; the measurement was repeated more than once until the deviation between two observations would not exceed one tenth of a pulse beat [2].

After this he measured the times needed to cover other distances on the board. He compared these with the time he had measured for the whole descent. Rolling one quarter of the length, for example, took exactly half the time of the whole length, as illustrated in Figure 4.1. "[I]n such experiments, repeated a full hundred times, we always found that the spaces traversed were to each other as the squares of the times, and this was true for all inclinations of the plane" [2]. He had obtained Equation 4.1.

To obtain an accurate mathematical law he had to measure time with sufficient precision. Since there were no mechanical clocks in Galileo's day, this was a crucial part of the development of the experiment. He describes a simple timekeeper in his book, built from a large vessel of water. A small pipe was soldered to its bottom to produce a thin jet of water that could be collected in a small glass during the time of each descent. After the descent he weighed the water on a "very accurate balance". The ratios of these weights, he said, gave the ratios of the times, "with such accuracy that there was no appreciable discrepancy in the results, although the operation was repeated many, many times" [2].

This experiment was groundbreaking in more than one way. Scientists had tried to understand velocity in terms of distances, but Galileo realized that it was more productive

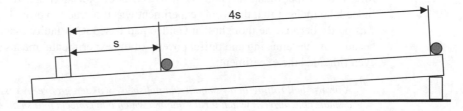

Figure 4.1
Schematic description of the experiment on the inclined plane. At the end of the slope is a wooden block. The two balls represent two starting points for the descent, and the distance *s* is one fourth of the total length. Galileo found that it took the ball exactly half the time to cover this distance as compared to the whole length.

to study it in terms of time. This made it possible for him to introduce the concept of acceleration. The experiment also marks the beginning of modern experimental science by being the first systematically planned experiment that resulted in a precise mathematical law [1]. In going from passive observation of the heavens to active experimentation, science had moved from being a spectator to taking the director's seat. Instead of sitting back and enjoying the show, it set the stage for Nature's performances and took command of the events on it.

The law of free fall made it possible for Galileo to establish that projectiles follow parabolic trajectories. As described in the previous chapter, this was the trajectory that Newton identified with the orbit of the moon when he merged the celestial and terrestrial mechanics. Galileo simply divided the motion of a projectile into a horizontal part and a vertical part. The horizontal motion was unimpeded and uniform, while the vertical motion obeyed the law of free fall. The result was a parabola. This meant that there was no difference between Aristotle's violent, upward motion and the natural, downward motion. They were one and the same motion, described by a single law.

Galileo's experiment on the inclined plane has passed into history as a turning point in science but questions have been raised about it. Was it really an experiment? Some have claimed that it cannot have been more than a physical demonstration of the law of free fall, which he somehow must have obtained by other means. Did Galileo really have the precise experimental means required to determine his accurate law?

4.2   Experiment or Hoax?

Marin Mersenne, a contemporary of Galileo, was a friar and scientist who corresponded regularly with many of the leading scientists of his day. He had seen references to the experiment on the inclined plane already before the publication of *Two New Sciences* and had tried to perform it himself. Although Galileo had mentioned the experiment in other places, he had never given a description as detailed as that in his last book. We may even suspect that the detailed description there was an answer to Mersenne's criticism. As Mersenne had not succeeded in reproducing the experiment, he regarded it as incapable of founding a science. He even doubted that Galileo had performed it at all [3].

Perhaps inspired by Mersenne's failed attempts, the philosopher Alexandre Koyré provided outspoken criticism of Galileo's experiment in a 1953 paper [4]. Koyré was an eager advocate of the view that theory must precede observation in science and, judging from the paper, he even seems to have thought that empirical investigation was less dignified than theoretical speculation. Galileo's water clock was too primitive to have played a central role in the experiment, he thought. It lacked both precision and theoretical refinement. Indeed, he thought that the whole experiment was too crude to provide useful results. This was partly because he thought that Galileo had based it on faulty assumptions, and partly because of the "amazing and pitiful poverty of the experimental means at his disposal". He even ridiculed the experiment:

> *A bronze ball rolling in a "smooth and polished" wooden groove! A vessel of water with a small hole through which it runs out and which one collects in a small glass in order to weigh it afterwards and thus measure the times of descent [...]: what an accumulation of sources of error and inexactitude!*
>
> *It is obvious that the Galilean experiments are completely worthless: the very perfection of their results is a rigorous proof of their incorrectness [4].*

In other words, he accuses Galileo of academic dishonesty – of fabricating data to confirm a mathematical relationship that he had obtained by means other than experimentation. He also declares that this must be the reason why Galileo does not specify a concrete value for the acceleration of gravity.

Koyré points out that it is wrong to assume that a ball rolling down an inclined plane is equivalent to a body gliding down the plane without friction. Due to the ball's rotational inertia, the rotation requires energy and this affects the speed. His most serious criticism, however, is the assertion that Galileo lacked the means to measure time with the required precision. Rather than investigating the water clock Koyré seeks to prove his point by explaining how difficult it is to measure time with what he considered to be a much more refined instrument: a pendulum.

Galileo himself had established that a pendulum of a certain length has a constant period of oscillation, meaning that a cycle of back and forth swings always takes the same time. This fact is said to have dawned on Galileo while looking at a swinging candelabra in the cathedral of Pisa and timing the motion with his own pulse. It is also known that he established it by experimenting with pendulums that had the same length but different weights. Despite this, Koyré emphasizes that Galileo had arrived at the result "first and foremost by hard mathematical thinking". In his eyes this seemed to make the pendulum a more worthy timekeeper than the water clock, and he thereby thought it rather strange that Galileo had not used it in his experiments with the inclined plane. The constant period made the pendulum a very suitable technology for timekeeping.

Father Mersenne had used a pendulum to measure time in his own experiments with free fall. According to him, the period of his pendulum had been exactly one second. When comparing his times of free fall with times measured by Galileo, Mersenne found large differences in the absolute numbers. Mersenne's measurements, which Koyré describe as progress compared to those of Galileo, also implied another, rather counter-intuitive result. Mersenne was forced to admit that the bob of the pendulum seemed to travel from its turning point to the bottom, along the curved path in Figure 4.2, in the same time that it would cover the same height vertically in free fall. Clearly, there must have been something wrong with these measurements. And using a pendulum to measure falling times in absolute units is challenging, to say the least. Those who have been working in a laboratory will have noticed that some people have a greater aptitude than others for standing up to such

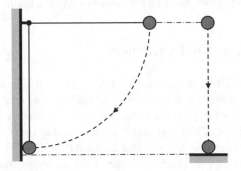

Figure 4.2
Mersenne found that the bob of the pendulum fell to its bottom position (left) in the same time that it would cover the same height in free fall (right). Something was obviously wrong with his measurements.

challenges. Koyré freely admits that Mersenne does not seem to have been one of them but he chooses not to pay this fact much attention. His preferred conclusion is rather that "precision couldn't be achieved in science and that its results were only approximately valid" [4]. If time could not be measured with a refined pendulum, it certainly could not be measured with a primitive water clock.

Koyré goes on to describe the work of the Jesuit scientist Riccioli, also known for being a persistent anti-Copernican (presumably, he may have felt compelled to hold that view). Needing an absolute timekeeper to determine the rate of acceleration in free fall he tried to build a pendulum with a period of exactly one second. Using a water-glass, he was able to confirm Galileo's finding that the period of the pendulum was constant. To calibrate the period, however, he needed to use the only exact time reference there was: the motion of the celestial bodies. First he used a sundial to count the oscillations for six consecutive hours. The result was disastrous, so he built a new pendulum and counted its beats for twenty-four hours, from noon to noon. Still dissatisfied, he adjusted the length of the pendulum and made a new attempt. To increase the measurement precision he now started counting when a certain star passed the meridian line and continued until it passed the line again, the following night. As the period still was not right, he made several new attempts. Despite heroic efforts, wearing down the patience of his assistants, he never managed to get the period right. Measuring time accurately with a pendulum was simply very difficult.

To finish the story off, Koyré turns his attention to Christiaan Huygens, the famed inventor of the pendulum clock. The work leading up to this invention included series of experiments but Koyré emphasizes the theoretical aspects. Huygens' clock is important, he says, because "it is the result not of empirical trial and error, but of careful and subtle theoretical investigation of the mathematical structure of circular and oscillatory motions". With his clock, Huygens could determine the exact period of a pendulum that he had used in his own experiments. On the other hand, Huygens did not have to measure the rate of acceleration in free fall, because through his work he had found a formula that related the period of a pendulum to its length and the acceleration of gravity, g. Using a pendulum and the clock he had invented, Huygens could thereby determine the value of g that is still accepted. This is what Koyré perceived as the final purpose of the Galilean experiment.

In Koyré's eyes this closed the circle. Through Huygens' theoretical ingenuity there was, at last, a timekeeper with sufficient precision to determine the rate of acceleration in free fall. The argument is wrapped up in the following moral: "not only are good experiments based upon theory, but even the means to perform them are nothing else than theory incarnate" [4].

4.3   Reconstructing the Experiment

Koyré leaves us with a couple of central questions. Could Galileo have conducted the experiment with his limited means? And was the aim really to measure the acceleration of gravity? In response to the criticism, Thomas Settle reconstructed the Galilean experiment in 1961. Then a graduate student in the history of science, he staged the experiment in the common living room he shared with other students [1]. The purpose was to get a better appreciation for some of the problems that Galileo had faced. Both the equipment and the procedures were kept as close as possible to Galileo's description in *Two New Sciences*. It turned out that he could easily reproduce the results with the precision claimed by Galileo [3].

Settle points out that Koyré had fundamentally misunderstood the purpose of the experiment. It was true that Galileo's apparatus was not capable of establishing the gravitational constant, but he was not trying to do that. He was interested in the ratios given in Equation 4.1. He was thereby not dependent on measuring time in standard units and could be completely arbitrary in his choice of measures. Furthermore, there had been little justice in Koyré's criticism of Galileo failing to account for the rotational inertia of the ball. Not only did this problem not exist in his mind, in this experiment it was even irrelevant. As the factor that accounts for rotational inertia is a constant, it does not affect the proportionalities of the law. Taking the ratios in Equation 4.1, the rotational inertia of the ball cancels out of the equation [3].

His reconstruction is interesting from many points of view. One is to see that it was technically feasible to establish the law of free fall from the experiment. The actual performance of an experiment also brings out many more of its dimensions than reading about it. Conceiving an experiment is only part of the effort. A number of practical and conceptual problems must be solved to bring it to full maturity. For example, as we noted in connection with Mersenne's measurements, the operator is clearly an integral part of the experiment. Settle notices the importance of being allowed a few practice runs before each test to obtain consistent results. It would be valuable, he writes, to have a more detailed account both of Galileo's thinking and practical work around the experiment:

> There is [. . .] a fascinating and vastly important body of knowledge concealed in the "conceiving" and "bringing to maturity" of both the theoretical and empirical aspects of this experimentation [. . .] For each step of original work we would like to know the mistakes and dead ends, the contributions and limitations of the existing technology and mathematics, the many conceptual aids as well as hindrances inherited from his contemporaries, and the nature and significance of his own predispositions.

Settle did his best to choose equipment and procedures that were no better than those available to Galileo. He used an 18-foot pine plank and cut a $\frac{1}{4}$-inch groove in it with a circular saw. After sanding it he applied wood filler and rubbed the surface with wax to make the edges hard and smooth. Although there were irregularities where knots or the grain crossed the groove, he made no further attempts to make the edges exactly parallel. Instead of Galileo's bronze ball he used a standard billiard ball and a steel ball. An ordinary flowerpot was used as a water container for the timekeeper. He threaded a small glass pipe through its bottom hole for the outflow. Its upper end was positioned high enough in the pot for him to cover it easily with a finger while his palm rested on the rim of the pot. Lifting the finger, the water flowed out into a graduated cylinder under the pot and was "weighed" by reading its volume in milliliters. For each run, Settle placed a wooden block at a predetermined distance down the slope of the plank and filled the pot with water. He simultaneously released the ball and lifted his finger from the pipe. At the sound of the ball striking the block he replaced the finger and the "time" of the descent was read in milliliters on the graduated cylinder. He estimated that this method for measuring the time was about five times as crude as the one available to Galileo.

The time measurements were the most difficult part of the experiment. A uniform water flow had to be kept at least for the duration of the longest readings. He also found it necessary to practice releasing the ball and the water flow at the same time, and stopping the flow immediately at the strike of the ball. The operator "must spend time getting the feel of the equipment, the rhythm of the experiment", he said. Sticking to techniques available in Galileo's time, he confirmed that the water flow was uniform by timing it with a pendulum. He made the pendulum from the billiard ball and kept the length a little less than a meter. This would make it beat at about the same rate as the pulse, which was appropriate since Galileo stated his precision in terms of "pulse beats".

Watching the shadow of the bob against vertically lined paper he could accurately lift and reset his finger on the pipe at the end of a beat. This procedure showed that the water flow was indeed constant within one tenth of a second. When rolling the ball he took several readings for each distance, just as Galileo had described, and sight-averaged them. (The technique of calculating means to get more precise values than single measurements was not established in Galileo's days.) With this simple method, even the largest deviation from the theoretically expected value was less than a tenth of a second.

Settle emphasized the simplicity and ease with which his results were obtained. He said that he maintained a "deliberately cavalier attitude towards procedures and measures" because he "wanted to give 'error and inexactitude' every reasonable chance to accumulate. And yet they did not" [3].

Not only did Settle highlight Koyré's misunderstanding of the purpose of the experiment. The reconstruction also showed that it was technically feasible for Galileo to make the measurements. But one central question remained unanswered. Was it an experiment? Historians of science have regarded the inclined plane as a way for Galileo to demonstrate the truth of a law that he had obtained by abstract reasoning, not by explorative tests. As we will see, a page in Galileo's notebook from 1604 casts some light over this question.

4.4 Getting the Swing of Things

Measuring time makes us think of accurate clocks and precisely defined units like seconds. Galileo was thinking in terms of ratios and, for him, it was sufficient to be able to divide time into equal intervals. Stillman Drake, Galileo's biographer, thought that he was able to do that using music.

As we have mentioned, Galileo composed music and is said to have played the lute quite well. Musicians are actually able to keep the time quite accurately for long periods while playing. They do this without external aids and without thinking of standard units like seconds. An internal rhythm allows them to keep an even beat, and to divide that beat in half again and again with a precision rivaling that of any mechanical instrument. Drake notes that, if the cymbalist in an orchestra were to miss his entry even by a 64th note, everyone in the audience would immediately notice it. For a musical person it should be perfectly feasible to accurately divide a period into equal intervals without any external timekeeper [5].

In the 1970s, Drake studied a previously unpublished page of Galileo's working notes. It contained lists of figures scribbled on a page. He says that it looked singularly unpromising at first glance but he slowly realized that it recorded an actual experiment. There were three columns on the page, which are reproduced in Table 4.1. The second column seemed to be an index, containing the integers from one to eight. The first column contained the squares of these integers, but it differed from the others by being written with a different pen and ink. Drake concluded that it must have been added after the other two. The figures in the third column very nearly represented the distances covered by a ball rolling down an inclined plane at the end of eight equal times. Drake points out that these distances

Table 4.1 Figures entered in Galileo's notebook, recording an actual experiment. The third column very nearly represents the distances covered by a ball rolling down an inclined plane at the end of eight equal times. The second column is an index. The first column was written after the others, in different pen and ink, and contains the squares of the figures in the second column.

1	1	33
4	2	130 −
9	3	298 +
16	4	526 +
25	5	824
36	6	1192 −
49	7	1620
64	8	2104

must have been measured, because they were not exact multiples of the squares in the first column. Also, they could not have been measured with the water clock, as described by Galileo in the published version of the experiment. That would require dividing a single time into eight equal intervals, which is impracticable using a constant flow of water. It could have been done using a pendulum, but keeping track of the motions of the ball and bob at the same time would have been challenging. Another possibility is to use the sound made by water droplets dripping into a container at a constant rate, but also this requires two external sensory impressions to be coordinated. Drake believed the experiment to have been much simpler and more elegant [5].

Being a competent lute player, there is little doubt that Galileo could keep a precise beat. If the ball could be made to produce sounds during its descent, they could be used to divide the time into equal intervals. One way to do this is to attach adjustable frets to the plane. Unlike modern string instruments, which have fixed frets, the lutes that Galileo was familiar with would have movable frets around their necks. Attaching such frets to the plane, the ball would make an audible bump each time it passed over one. In his reconstruction of the experiment, Drake could easily find the approximate positions of the frets by firmly singing a slow march while releasing the ball. In just three or four runs he was able to make eight chalk marks on the board that represented the beats of the song. Instead of the gut frets that Galileo would have used, he put rubber bands around the plane at these positions, as indicated in Figure 4.3. It was then easy to adjust the locations of the frets during a few runs to obtain an even beat. The accuracy of the results was better than a 64th of a second.

Drake's reconstruction led him to suspect that the occasional plus and minus signs after figures in the third column of Table 4.1 indicated that Galileo thought that those sounds seemed a little early or late, but not enough to require further adjustments. It also allowed

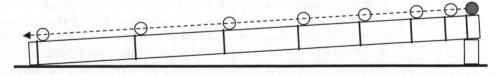

Figure 4.3
Schematic description of Drake's reproduction of Galileo's experiment. Placing rubber bands on the board, the ball makes audible bumps when passing them. In a few runs the locations of the bands can be adjusted to produce an even beat.

him to find a physical explanation for a certain discrepancy in Galileo's data. His second fret seemed to be placed a little earlier than theory would predict. Drake saw that the ball was delayed in climbing the first fret, since it moved slower in the beginning of its descent. To obtain an even beat the second fret had to be moved closer than theory predicted, otherwise the second time would not be equal to the first. Once the ball was underway it passed the other frets without effort.

With this ingeniously simple method it was obviously possible to divide the descent of the ball into equal time intervals more accurately than any water timer could measure, and without having to rely on any other external means. The figures in Table 4.1 also imply another important conclusion: Galileo did not yet know the times-squared law when he performed the experiment. If he did, it would have been pointless to measure the distances. A single run would have been enough to confirm the law and there would have been nothing to write down. Calculations on the manuscript page along with evidence that the first column was written after the third indicated that the measurements led Galileo to the law of free fall, not the other way around. The experiment was an exploration of nature, an attempt to seek an unknown rule from measurements. Drake also pointed out that if his reconstruction was correct it would surely explain why Galileo never published the experiment. Even in his days it would be foolish to state that he had arrived at a precise mathematical law by humming a song while watching a ball roll down a plank [5].

Galileo's notes reveal that he probably performed the experiment essentially as Drake describes it. Regardless of this, the reconstruction teaches us another important thing about the experiment: there are different ways to carry it out. Even without a sophisticated timekeeper, there is more than one way to obtain Galileo's law with the required precision, using little more than an inclined plane and a ball.

4.5   The Message from the Plane

We began this chapter with a number of questions. One of them has not yet been explicitly answered: what is an experiment? Galileo saw an experiment as a question put before Nature. The secret of a great experiment lies more in how well that question is put than in the measurement technology used. For instance, an accurate timekeeper would add very little value to Aristotle's experiments with free fall in water because he posed the question in a way that muddled the problem. Precision can make it easier to interpret the results of an experiment that is well conceived, but if we want Nature to answer our questions a keen eye and a bit of imagination are more valuable assets than a precise instrument.

It would be preposterous to judge the scientific quality of an experiment by the sophistication of the measuring instrument used. In that case, weighing your fruit on the electronic scales at the supermarket would be far more valuable than many groundbreaking experiments in the history of science. The theoretical knowledge embodied in such scales is, after all, considerable. Since scientists seek to understand the world it is important for them to recognize trends and regularities. Scientific progress does not mainly lie in the acquisition of accurate numbers; it lies in the identification of the patterns behind the numbers. Galileo's experiment is important because it shows how a simple setup can unveil a distinct pattern in Nature.

Koyré seems to have been more interested in his own agenda than in this aspect of the experiment. In his mind, he had reduced the purpose of the experiment to the measurement of a constant, but Galileo was doing something more complex and interesting than that. He was trying to find out how acceleration occurs. We should not confuse experiments with measurements. Measurements are just structured ways to acquire data and they are made all across society, from research laboratories to the fruit market. Measurements are not a unique characteristic of the experimental method, because passive observers also measure things. What makes experimenters unique is that they *interact* with what they measure. They create specific conditions to answer specific questions.

Settle showed that there is more to a good experimenter than the ability to implement theoretical knowledge in a laboratory setup. The will and ability to handle the "nuts and bolts" of the setup is often crucial for success. Theoretical considerations do not necessarily dictate which approach to a problem is most suitable; practical considerations are equally important. An experiment is not theory incarnate.

Drake showed us that it is very likely that Galileo used the inclined plane to make a genuine experiment, rather than just a demonstration. He also showed that there is much more to science than precision in measurement and calculation. A successful experimenter needs the ability both to ask meaningful questions and to find useful means to answer them. Though precise measurements and calculations are significant features of any exact science, a creative mind is equally important. Experimenters are like artists – they must work with mind, hand and eye, and find a harmony between them.

It must be said that experimentation is a creative activity. The inclined plane was not a given. Galileo had to find a way of exploring his problem that made the investigation possible, and he created the plane to this end. The French author Émile Zola said that art is a corner of creation seen through a temperament. Likewise, science can be said to be a corner of reality seen through an intellect. The difference is that science must stand up to scrutiny. Where art is a subjective reflection, science must give an objective account of reality. But that is true of the theories that are the end result of scientific enquiry. The path to them is often hilly and covered with brushwood; clearing a way through the thicket may require considerable creativity.

Experimenters do not settle with collecting numbers. They create unique conditions to explore the world around them. At the end of the day, experimenters see the world more through an artist's eye than through the eye of an accountant.

4.6 Summary

- Experimentation is a more efficient way to obtain information than passive observation. Where the passive observer may have to deal with large amounts of irrelevant information, the experimenter creates conditions that are relevant to a specific problem.
- As scientists aim to understand nature, it is often more important for them to recognize patterns than to be able to measure with precision.
- Experimentation is not to be confused with measurement. An experiment is an interactive exploration of nature. A measurement is an organized way of obtaining data, but may still constitute a passive form of observation.
- Experimenters often need the ability to conceptualize a problem. The inclined plane is a successful example of such conceptual thinking, allowing Galileo to study free fall using the technology of his time. Aristotle's method of slowing objects down in water is a less successful example. The problem must be simplified without disturbing or removing its essential features.
- Some important experiments test theories, but not all. Experiments may be pure empirical explorations, driven by curiosity alone.

References

1. Crease, R.P. (2004) *The Prism and the Pendulum*, Random House, New York.
2. Galilei, G. (1914) *Dialogues Concerning Two New Sciences* (H. Crew and A. de Salvio, trans), Dover Publications, New York.
3. Settle, T.B. (1961) An Experiment in the History of Science. *Science*, **133**(3445), 19–23.
4. Koyré, A. (1953) An Experiment in Measurement. *Proceedings of the American Philosophical Society*, **97**(2), 222–237.
5. Drake, S. (1975) The Role of Music in Galileo's Experiments. *Scientific American*, **232**, 98–104.

5 Scientists, Engineers and Other Poets

Why should there be the *method of science? There is not just one way to build a house, or even to grow tomatoes. We should not expect something as motley as the growth of knowledge to be strapped to one methodology.*

—Ian Hacking

One reason why many Ph.D. students find it difficult to connect the philosophy of science to their research is that philosophers like to describe the scientific method using examples from the natural sciences, especially physics. They discuss predictive theories expressed in mathematical terms, which describe the most fundamental mechanisms in the world. In many sciences, theories are rarely mathematical or predictive but still have great explanatory power. Evolution by natural selection is just one example. Students in engineering science do not always work with basic mechanisms and some of their research lies very close to technical development. This makes it difficult for many students in applied research fields to relate to scientific method as it is described in textbooks. To amend the situation, general characteristics of research will be given here that work both in basic and applied fields. I will also make the point that there is a distinct difference between development work and academic research. It is important to focus on research during the research studies, as this is what research students are supposed to learn.

The long-term goal of science is to explain and understand what happens in the world. This understanding is embodied in scientific theories. If we are to contribute to theory it helps if we understand what theories are and how they are related to reality. We will take a look at this topic and then turn our attention to something that is almost always overlooked in texts on scientific method: creativity. In some respects, the difference between scientists, engineers and poets is smaller than many of us think.

5.1 Research and Development

Many Ph.D. students carry out research in applied fields. They may share techniques, methods and practical tasks with people who develop products in industry. The Ph.D. student is usually awarded an academic title after a few years, but the engineer is not. If that title is to have any real value there must be a crucial difference between what they do.

In this book, the term research refers to scientific research – not to the work carried out in Research & Development organizations in industry. I will make the point that the difference between research and development lies in the purpose and outcome of the tasks, rather than in individual techniques and tasks. We may get an idea of what characterizes development

Experiment!: Planning, Implementing and Interpreting, First Edition. Öivind Andersson.
© 2012 John Wiley & Sons, Ltd. Published 2012 by John Wiley & Sons, Ltd.

work by looking at how the Accreditation Board of Engineering and Technology (ABET) in the United States defines the engineering profession:

> *Engineering is the profession in which a knowledge of the mathematical and natural sciences, gained by study, experience, and practice, is applied with judgment to develop ways to utilize, economically, the materials and forces of nature for the benefit of mankind.*

This, indeed, sounds like a respectable profession, but is engineering really the only occupation characterized by these things? Scientists, too, use mathematics and science with judgment and their work is funded because society expects to benefit from it in the long run. Let us keep this definition of engineering in the back of our minds and return to it later.

Drawing a line between engineering and science has nothing to do with how we value these activities. Our aim is simply to find the unique characteristics of research. If we cannot distinguish the scientific method from other methods, it is meaningless to put a name on it. It is especially important for researchers who teach and examine Ph.D. students to know what it means to "master the scientific method", because this is an important learning outcome of the research education. If we cannot define it, how do we know if a student has learned it?

An interesting approach to the problem is provided by Mischke [1], who attempts to draw a line between science and engineering by pointing out that they are governed by two separate types of interest. Figure 5.1 shows some essential features that define these interests. There is a *system*, which is governed by the *Laws of Nature*. These laws determine the *output* that will result from a specific *input* to the system. We could, in principle, take two types of interest in these features. Our focus is either on the system itself or on the laws that govern it. Scientific research uses the system, the input and the output to find the governing *laws*. Engineering uses knowledge of the laws to find a *system* that will produce a desired output from a certain input [1]. It should be noted that the system does not have to be a physical object. Both scientists and engineers may be interested in non-physical systems like methods, computer programs and other processes, or even behaviors.

It is easy to find examples that match these criteria. For instance, using the laws of thermodynamics to devise an engine (system) that meets a certain fuel consumption target (output) in a specified test cycle (input) would clearly be engineering. Rolling a ball down an inclined plane (system) and measuring the times (output) required to cover predetermined distances (input) in order to establish Galileo's law would clearly be research, at least if

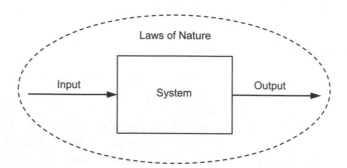

Figure 5.1
Four common features of research and engineering activities. There is a system that is governed by the laws of nature. Input into the system results in an output. Research uses the system, input and output to find the laws. Engineering uses the laws, input and output to devise the system. Adapted from Iowa State Press © 1980.

it was done for the first time. On the other hand, it is equally easy to find examples of acclaimed researchers who have won their fame through activities that, by these criteria, must be considered to be engineering. Robert Millikan, whom we are going to meet in the next chapter, is one example. He was awarded the Nobel Prize for determining the charge of the electron. To do that he devised an ingenious apparatus where charged oil droplets were suspended in an electric field. Balancing the downward force of gravity by an upward electric force he could make the charged droplet hover. This made it possible to determine the droplet's charge, which turned out to always be a multiple of a certain number – the electron's charge. Clearly, to conceive this apparatus (system) he had to use his knowledge of the laws of gravity and electricity, and his knowledge of how these would produce a desired output (hovering) from a certain input (charges). And Millikan is not alone. The history of science is full of similar examples of "engineering". Since scientists apparently do a great deal of this we should perhaps use a more general word for engineering, such as development, because all who develop things are clearly not engineers. In any case, research tasks cannot be defined in terms of their difference to engineering or development, since even basic research may involve such tasks. Mischke's criteria do not work if we apply them to individual tasks.

As previously stated, science aims to explain what happens under various conditions: the goal is general knowledge about the world. Another possible approach to our question is, therefore, to ask how general or specific the knowledge that results from certain activities is. As basic research explores the fundamental properties of nature it aims at the most general knowledge we can have about the world. Engineers also seek knowledge but of a more specific kind. In engineering, specific systems are developed to meet specific needs. Experiments may be used to find out how a system behaves but the goal is not to uncover all its fundamental properties. Engineers only need enough knowledge to be sure that the system will perform its intended function. Applied research finds itself somewhere between these two extremes. Here, researchers may work with quite specific applications, such as diesel engines, water purification, or fire safety. They may be interested in developing new methods or measuring techniques that are relevant to those fields. They may also be interested in testing the implications, limits or even the validity of established theories within their fields. Or they may be less focused on theories and more interested in explorative tests, stretching the limits of what is currently possible within their area of application. Regardless of their focus it is important to note that, although they are interested in a specific application, they seek to gain general knowledge and understanding at some level. Even if their work does not generate new theories, the results must still have a general relevance within their research community. If results obtained in one laboratory cannot be applied in other laboratories, they have no scientific relevance.

The distinction between research and development is a matter of the long-term aim of the activities. All researchers, even those in fundamental fields, must rely on specific observations made in specific situations, but they use those observations to develop *general* knowledge. Engineers, on the other hand, are interested in *specific* knowledge. Even when they apply general theories they do it to find out if a particular system works well enough or how it can be made to perform better.

In summary, if we want to discern research from other activities, it is too superficial to look at practical tasks. We need, instead, to look to the purposes of the tasks. Returning to ABET's definition of engineering we note that it does not only state what engineers do. It states the *purpose* of the profession, which is "to develop ways to utilize, economically, the materials and forces of nature for the benefit of mankind". Scientists are not mainly concerned with what is economically and technically viable but with providing a knowledge base for others to build on. Looking to the purposes rather than the tasks we also see that Mischke's demarcation between science and engineering remains useful. Even though the individual tasks of scientists and engineers may be similar, the *purpose* of science is to understand the laws and the *purpose* of engineering is to devise the system.

> **Exercise 5.1:** Make a copy of Figure 5.1 and apply it to your own research project. Specify the system, input, output and governing laws that you are interested in. Are you doing research or development?

5.2 Characteristics of Research

Being able to spot the difference between development and research is one thing. Being able to state unique, defining characteristics of research is another. We will take a look at two attempts to do this and discuss how well they capture the essential features of scientific work.

In a popular book on research methods, Leedy and Ormrod [2] explain that, although research activities may differ substantially in complexity and duration, they typically have the following eight distinctive characteristics in common:

- Research originates with a question or problem.
- Research requires the clear articulation of a goal.
- Research requires a specific plan for its realization.
- Researchers usually divide the principal problem into several, more manageable sub-problems. Solving these will solve the main problem.
- Research is guided by the specific problem, question, or hypothesis that it originates from.
- Researchers accept certain critical assumptions, which serve as a foundation for their analysis.
- Research requires collection and interpretation of data.
- Research is cyclical in nature. New problems are often identified during scientific study and to solve these the process must begin anew.

Most researchers probably agree that this description captures some important features of research. The question is if they are sufficiently unique to determine if a certain set of activities is research or not. If your car does not start one morning, the troubleshooting process that you go through to fix the problem is likely to be characterized by the same eight features. These features are also applied in customer surveys and product development projects. If we go to the Nobel Prize ceremony, these features are probably common to the research programs that are rewarded with the prize but, in fact, they also apply to the work of the chefs in charge of composing the menu for the Nobel banquet. These chefs begin with the problem of how to create a gastronomic experience that suits the occasion. Second, their goal must be clearly stated. Otherwise they cannot plan their work, which is the third point. Fourth, they certainly divide the problem into subtasks, such as composing the individual dishes or choosing wines to go with them. Fifth, it would be rather surprising if their work were not guided by the main problem of composing a gala dinner menu. Sixth, when composing a menu, chefs accept certain critical assumptions. For instance, they assume that certain combinations of flavors are more pleasing to the palate than others, and that certain dishes should have certain qualities and be presented in a certain order with respect to each other. Seventh, we hope that the chefs carefully choose good raw materials and that they critically taste their creations to make sure that they reach the desired result. This means also that chefs collect and interpret data to solve their tasks. Finally, composing menus is a cyclic activity, since the work with one menu is likely to spur new ideas, which may serve as inspiration for the next menu. In summary, the features of Leedy and Ormrod may characterize research but they are not unique to research, since they apply also to other activities.

Phillips and Pugh [3] make a more promising attempt to define research. They restrict themselves to three characteristics of good research. The first is that research is based on an *open system of thought*. This means that there are no hidden agendas to keep in mind. As a researcher, you are both entitled and expected to question anything. We can see that this sets science apart from other modes of thinking, such as politics, management, religion or marketing, where mission may take precedence over evidence. In science, the testing and criticism of other researchers' work are ends in themselves and are important for the development of knowledge. The second characteristic is that researchers *examine data critically*. This could be seen as a consequence of the first point but it is so essential that Phillips and Pugh choose to list it separately. It is probably the single most important element distinguishing science from other approaches. A researcher's immediate response to provocative statements like "Women make less efficient managers" is not to agree or disagree but to ask: "What is your evidence?" Researchers constantly ask if they got the facts right, if they could get better data or interpret the results differently. Outside research, people tend to be impatient with such questions. Researchers, on the other hand, must exert themselves to obtain systematic, reliable and valid data because they aim to understand and interpret. The third and final characteristic is that researchers *generalize and specify the limits on their generalizations*. Valid generalizations make it possible to apply knowledge in a wide variety of appropriate situations, which is the very idea of science. Generalizations are best established through development of explanatory theory. Relating observations to theory is indeed one of the things that turns mere data collection into research [3].

The three characteristics proposed by Phillips and Pugh are useful because, together, they define *unique* aspects of research. There are other characteristics too, such as the ones proposed by Leedy and Ormrod, and we could add still others to the list. For example, research is a *systematic* search for knowledge, meaning that it is based on methodology. We could say that research is a *collective* process, considering the importance of the scientific community. We could also say that research is a *creative* process, of which more will be said later in this chapter. However important these aspects may be, they are not unique to research. But if we are involved in an activity that is based on an open system of thought, where we have to examine data critically, and if the purpose is to extract general meaning from them, we can be quite certain that it is research we are doing. Figure 5.2 shows this "trinity" of research characteristics graphically.

Exercise 5.2: Find one or several published research papers and examine them with the three characteristics of Phillips and Pugh in mind. Are the criteria fulfilled? (This exercise is useful to carry out in a group with subsequent discussion.)

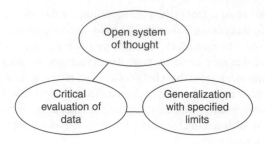

Figure 5.2
Three characteristics that are unique to research if they occur together.

Exercise 5.3: Examine your own research project with the three criteria of Phillips and Pugh in mind. Are they fulfilled?

If there is one thing that could be added to the list of Phillips and Pugh, it is this: research is a *self-corrective process*. It is arguably a consequence of the other criteria but it deserves to be mentioned separately. The peer review process of research papers before they are published is one expression of this effort of self-correction. We have also seen that even established theories are discarded if they turn out not to hold water. It does not matter how elegant they are or how fond we are of them. Researchers hold no sentimentality for faulty ideas. It is difficult to think of any other system of ideas that has this sort of self-review so centrally built into it. How often do we hear political ideologists, management consultants, religious leaders or marketing officials announce that central parts of their messages have been fundamentally wrong and must be revised in light of new evidence? Scientists may well make such announcements with enthusiasm, relieved that an obstacle to progress has been removed.

After this attempt to define research, let us turn to the knowledge that results from it. As Phillips and Pugh said, general knowledge is best established through development of explanatory theory. Although individual research activities may not be directly aimed at developing or testing theories, general theories are still the overall goal of science. If we are to learn how to carry out research we need to answer two central questions. What do we mean by theories, and how can we contribute to them?

5.3   Building Theories

Students in the physical sciences often think of theories as mathematical equations that sum up the essential characteristics of a certain class of phenomena. Are such laws the goal that we must aspire to if we want to contribute to theory? Molander [4] concisely defines theories as "systems of statements, some of which are regarded as laws, that describe and explain the phenomena within a certain area of investigation in an integrated and coherent way" (author's translation). By this definition the answer to our question is no, because statements are not necessarily made in the language of mathematics. It is, in fact, quite easy to find examples of important theories that are not mathematical, such as plate tectonics in geology or evolution in biology.

Theories make statements about what we are going to call *theoretical concepts*. These represent certain aspects of reality and are used to facilitate our thinking about the world. Theories can be seen as rules that apply to the concepts. Newton's third law, for example, states that "the forces of two bodies on each other are always equal and in opposite directions". This is clearly a strict and general rule that applies to the concepts of forces and bodies. But we do not have to aspire to come up with something like Newton's laws to make valuable contributions to theory. We only have to contribute with good, general knowledge that in due course may be integrated into a greater, more general theory. The established theories that we have learned about in textbooks have not emerged from a vacuum. Most of them build on a multitude of contributions from various researchers. For example, both Galileo and Kepler contributed to the development of Newton's mechanics, and they both built on the work of others.

The building of theories begins by looking at the world and describing it. Organizing the characteristics of a phenomenon is a necessary preparation for trying to understand its underlying mechanisms. Theoretical physicist Richard Feynman uses Mendeleev's discovery of the periodic table of elements as a good example of this [5]. The table is based on the observation that some chemical elements behave more alike than others and that their properties repeat periodically as the atomic mass increases. The table does not explain why the pattern exists but it makes the pattern readily visible. The explanation for the periodic behavior came later with atomic theory.

According to Feynman, the search for a new theory begins by making a guess about the reason behind a certain phenomenon. The next step is to work out its various consequences to see if it tallies with reality: "If it disagrees with experiment it is wrong. In that simple statement is the key to science" [5]. This sounds very much like the hypothetico-deductive approach introduced in Chapter 2 but this, Feynman adds, will give us the wrong impression of science because

> [...] this is to put experiment into a rather weak position. In fact experimenters have a certain individual character. They like to do experiments even if nobody has guessed yet, and they very often do their experiments in a region in which people know the theorist has not made any guesses. [...] In this way experiment can produce unexpected results, and that starts us guessing again [5].

This description sums up the point that we made in Chapter 2 about science not being purely hypothetico-deductive: sometimes observation precedes theory and sometimes it is the other way around.

The aim of developing theories is to come as close as possible to a general understanding of the world. We wish to represent reality by ideas or rules that are so general that we may, in some cases, even call them Laws of Nature. This could be seen as a mapping process where theory is the map and the real world is the terrain to be surveyed. The process is sketched in Figure 5.3. Parts of reality that are not yet covered by any theory are white areas on the map. To expand our theoretical knowledge we need to know its boundaries. We also need strategies for moving them forward. In the hypothetico-deductive approach this is done by formulating hypotheses, which can be seen as theoretical outposts in the uncharted terrain. We collect data to see how well the hypotheses represent reality; a wide variety of observational support is generally required to incorporate an hypothesis into the body of established theory. But this gives a limited view of how theory grows. Previous examples in this book show that the growth is not necessarily an additive process where new parts are added to the existing ones. Sometimes we suspect that two areas of

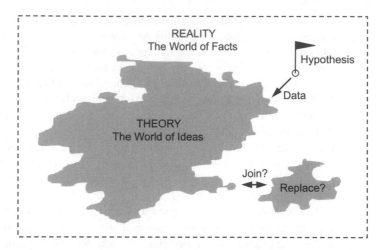

Figure 5.3
Development of theory can be seen as a mapping process. In the hypothetico-deductive approach new areas are charted by suggesting hypotheses and comparing them with data. Theory can also grow in other ways.

theory may be joined into a larger one, as when Maxwell joined electricity and magnetism into the unified electromagnetic theory that explained more. Older theories may also be replaced with new ones that explain more aspects of a given phenomenon.

Example 5.1: To illustrate the role of theoretical concepts in the development of theory, let us look at a phenomenon that has been known to humanity for a long time: fire. The complexity of combustion is daunting. In a flame, thousands of chemical reactions occur at the same time, often in a turbulent flow. The detailed description of both the turbulence and the chemistry goes beyond the state of the art of these sciences, and in combustion they are coupled in one problem: the turbulence affects the chemistry of the flame and vice versa. How do we even begin to formulate theories about something as complex as fire?

To illustrate this I will use a very brief (and simplified) example from my own field of expertise, diesel engines. As we said before, theories begin by looking at the world and describing what we see. When looking at a burning jet of diesel fuel, we may observe that it always burns as a *lifted flame*. The opposite of this would be an *attached flame*, such as the flame sitting on the wick of a candle or on a Bunsen burner. If you increase the gas flow sufficiently in a Bunsen burner, the flame will detach from the nozzle and become lifted. This happens when the flow speed at the nozzle exceeds the flame speed, which is the speed at which a flame propagates through a stationary gas mixture. Already at this point we have introduced the simple theoretical concepts of attached and lifted flames to describe a process that is fundamentally enormously complex.

In a lifted flame, the distance from the nozzle to the flame is called the *lift-off length*. This is another theoretical concept that is important for the understanding of diesel combustion. Combustion in diesel jets occurs in two stages [6], which are schematically described in Figure 5.4. Fuel flows into the combustion chamber from the nozzle at the left and starts to react at the lift-off position. At the center of the jet is a *premixed reaction zone* – another

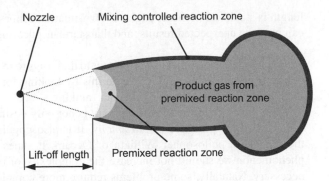

Figure 5.4
Schematic description of a burning diesel jet. Fuel is injected into the combustion chamber from the nozzle. Flame reactions commence at the lift-off position and occur in two steps. The mushroom shape arises from fluid phenomena.

theoretical concept used to describe the situation when fuel and oxidant are mixed prior to combustion. The product gas from this zone is surrounded by a second, *mixing-controlled reaction zone* at the periphery of the jet. "Mixing-controlled" refers to fuel and oxidant being mixed during the course of combustion. The reason for the two-stage process is that the premixed stage is *fuel rich* – there is not enough air present to completely oxidize the fuel. Partially oxidized products from this stage mix and react with fresh air at the jet periphery, giving rise to the mixing-controlled zone. We have now used four more theoretical concepts to concisely characterize some essential features of a process that, at first sight, seems too complex to describe at all.

After this sort of characterization we can try to explain interesting aspects using the concepts. For example, diesel engines are often connected with black smoke that mainly consists of soot formed during combustion. The *lift-off length* is believed to be important for understanding the formation of soot in the diesel jet. Soot can be seen as carbon atoms from the fuel that are not completely oxidized, and thereby lump together into particles. As the first, premixed stage of combustion is fuel rich, it produces large amounts of soot. The more air that is entrained in the diesel jet upstream of the lift-off position, the less fuel rich the premixed zone becomes and the less soot is formed. It is no use trying to entrain air into the jet downstream of the lift-off position, because it will be consumed in the mixing-controlled zone. Air entrainment prior to lift-off should, therefore, control the formation of soot.

It could seem like we now have a good understanding of the limiting factors behind soot formation in diesel jets. Theory says that the lift-off position is stabilized at a certain distance from the nozzle due to the balance between the flow speed and the flame speed. The flame speed is more or less constant, whereas the flow speed decreases with the distance from the nozzle. The lift-off length, in turn, determines how fuel rich the premixed zone becomes and this determines how much soot is formed. That is what we thought, but the situation is complicated by some recent experimental findings. These indicate that, under the conditions prevailing in diesel engines, the lift-off length is not stabilized through a balance between flow and flame speed. It rather seems to stabilize through a process that is more closely related to auto-ignition [7]. The old theory is thus not adequate under these conditions and, at present, we have no good theory available to explain how the lift-off length is stabilized in diesel engines. The situation echoes Feynman's words, "experiment can produce unexpected results, and that starts us guessing again". ■

The example shows that it is important to find ways to reduce the complexity of a problem before trying to understand it. We do this by looking for general patterns and describing them using concepts that can guide our thinking.

When it comes to theories, simplicity is not only desirable, it is a requirement. This is often called the principle of *parsimony*. It is also popularly called Occam's razor, after the mediaeval philosopher William of Occam. It states that when we try to explain a phenomenon we should not increase the complexity of our explanations beyond what is necessary. Naturally, some problems require more complex solutions than others but that is not a contradiction. The principle merely states that we should simplify as much as possible, not more. Parsimony is fundamentally a question of intellectual hygiene. We should never add unnecessary features to a theory, because complexity makes it difficult to see underlying patterns. The need to patch a theory up with various features is often a sign that something is wrong with the idea behind it.

We have already seen how the principle of parsimony works in connection with the development of the Copernican system in Chapter 3. This system was favored not because it was more accurate than the Ptolemaic one – which it initially was not – but because it was easier to understand. The problem was that it incorrectly represented the planetary orbits by circles. The first attempt to correct the prediction errors was to add epicycles to the circular orbits, which made the model less parsimonious. When Kepler realized that the orbits were ellipses he decreased the complexity again and improved the accuracy at the same time. In retrospect we see that this was a sign that he was closer to the truth.

What characterizes a good theory apart from simplicity? Good theories should of course be testable. If we cannot test their validity they tell us nothing about the world. They should also be both good and useful representations of reality. A good representation is general, meaning that it applies to a wide variety of conditions. For a predictive theory "good" also means accurate. Not all theories are mathematical, but in order to be predictive a theory must be expressed in mathematical form. The usefulness aspect has more to do with how easy the theory is to apply. For example, Newton's mechanics may not be quite as general as quantum mechanics but we still use them, for instance, to calculate the orbits of satellites. We could, in principle, calculate such orbits with quantum mechanics as well, but that would be much more complicated and less intuitive. Newton's theory is simply more useful for this purpose.

We can roughly divide theories into two groups. One of them is *explanatory theories*, which are capable of explaining a class of phenomena as consequences of a few fundamental principles. The other group is *phenomenological theories*. These treat relationships between observable phenomena without making assumptions about their underlying causes. Thermodynamics and optics are examples of phenomenological theories. Thermodynamics is a system of laws, such as Boyle's law which states that the pressure of a gas is inversely proportional to its volume in a system where the temperature and mass are kept constant. That law was established directly from observation and is purely descriptive. Phenomenological laws are common also in applied areas. It is possible, for example, for a medical researcher to establish from observation that a certain feature in an ECG (electrocardiograph) indicates a certain heart disorder without knowing exactly why. This would be a general rule that applies to a theoretical concept (the specific ECG feature) but in itself it does not explain the connection between the feature and the disorder. Such phenomenological laws must often be formulated before deeper, explanatory theories can be developed.

The best explanatory theories are based on a few simple and general rules from which phenomenological laws eventually can be *explained*. The kinetic theory of gases, for example, explains why temperatures, pressures and volumes of gases behave like they do by considering the gas to be a system of small particles in constant, random motion. Based on this and other assumptions, Boyle's law and other thermodynamic laws may be derived by deduction. Another example of a good explanatory theory is Maxwell's equations, from which the laws of optics can be derived.

The line between phenomenological and explanatory theory is not distinct. If we consider the picture of the burning diesel jet again, we noted that we do not fully understand why the lift-off length is established at a certain position. Based on a wide range of observations we may still formulate a phenomenological law that describes how the lift-off position is affected by various variables. Despite its phenomenological nature, this law will help us understand trends in soot formation, as described in the example. Even phenomenological laws thus have some explanatory power. In a similar manner, we could say also that explanatory theories like Maxwell's equations have phenomenological features. Despite them explaining everything from optics to radar as consequences of a general set of basic rules, we do not know the fundamental reason why the rules work. During the last century, physicists were working hard to unify the theory of electromagnetism with that of gravity to find a common cause for the two, just like Maxwell unified electricity and magnetism before them. But even if that had been achieved, the reason for the unified theory would still be unknown. The problem threatens to degenerate into an infinite regression of unknown causes. At one level or other, all theoretical laws become phenomenological.

As researchers we should not let ourselves be disheartened by the fact that we may never find the ultimate cause for everything. That would keep us from realizing that our most general theories actually constitute very powerful knowledge of Nature's machinery. The fact that we can identify such wide varieties of phenomena as the consequences of such a small number of general rules is quite awe inspiring. It is, for example, a direct, logical consequence of Newton's three laws of motion that distant galaxies are shaped as they are. At the same time these laws explain why your bread usually falls with the buttered side down when you drop it from your breakfast table. Neat, isn't it?

5.4   The Relationship between Theory and Reality

Philosophers of science often discuss whether theories and what they make statements about are "true" or not. Depending on if they are prone to answer yes or no to the question they divide into two camps: realists and anti-realists. In this context, covering both views from a philosophical standpoint would require too much space. Further reading is suggested at the end of the chapter but it is useful to discuss one aspect of the topic here.

Although most of us agree that there is an absolute reality, we must admit that there is some subjectivity in how we choose to look at it. Theories do not exist in nature. They are created by people and exist only in peoples' minds. Despite this, we hope that our theories and the theoretical concepts they build on closely correspond to reality. We cannot be absolutely sure that they do, because completely different theories sometimes make successful representations of the same part of reality. A good example of this is how Einstein and Newton treated gravity in completely different ways. Newton used the theoretical concept of force to describe it. In his theory, forces were propagated between bodies and acted instantly over vast distances. In Einstein's general relativity theory gravity is, instead, represented by a deformation of the four-dimensional fabric of the universe that he called spacetime. Massive objects distort spacetime and the trajectories of other objects moving in their vicinity thereby become curved. The distortion even affects the trajectories of particles without mass, such as photons. Einstein removed the concept of force from gravity and introduced the concept of spacetime, giving us a new, independent representation of the same part of reality. This raises a central question. If theoretical concepts are imaginary constructs, how can we claim that theories *explain* phenomena in the real world?

It is important not to understand the word imaginary as synonymous to make-believe. Theoretical concepts would be useless unless there was a reality behind them that gave them meaning. To clarify this point, let us consider the theoretical concept of seasons. If I were to say to my neighbor that springtime arrived early this year, he would of course know exactly what I meant. We both know that spring refers to a period of transition between colder and warmer weather that occurs every year (at least in temperate zones). We determine its occurrence by signs in nature such as the leafing and blooming of trees and plants. Meteorologists define seasons in terms of the average temperature over a month. Despite these concrete signs and strict definitions, and the fact that spring *feels* very real when a long winter finally yields, springtime is just an arbitrary theoretical concept invented to make our daily lives easier. There are, of course, real causes behind the transitions between the seasons. The most important one is the continuous and periodical variation in the direction of the earth's axis in relation to the sun, but there are also natural fluctuations in the weather systems and climate that affect the temperature. In reality there are no seasons, but only small, gradual variations in temperature. We have merely agreed to use the seasons as practical means to represent the *effects* of these changes. It is possible to use other representations instead. For instance, it is common to use the theoretical concept of months instead, which is based on other criteria. This general line of reasoning also applies to gravity. Newton's force of gravity is an abstract representation of reality that has certain useful features. It has physical meaning because forces can be measured. Einstein's spacetime also has physical meaning, since it predicted the unknown fact that gravity bends light. At the end of the day, it is very difficult for us to know if forces or spacetime are "true" representations of reality or if they are just useful concepts.

Despite this, we can still think of theories as knowledge of the world. If they are useful and general representations of reality we can use them to explain or predict what will happen in specific situations. We may even use them to discover formerly unknown aspects of reality, such as gravity bending light. If they were not "true" we would not be able to use them like that. Most people probably agree that Maxwell's electromagnetic theory constitutes true knowledge of nature, as it predicted the existence of radio waves and made radio communication possible. We use radios every day and know that the theory works. The electromagnetic wave is a theoretical concept that may seem abstract, but it clearly represents knowledge of the world.

To summarize, theories make general statements about theoretical concepts. These may be more or less abstract, but they represent measurable aspects of the real world.

In this and the last chapter we have seen that the successful development of both theory and experiments requires useful concepts that are products of our subjective view of a problem. This means that science is a creative activity. As this aspect of scientific thinking rarely is treated, I would like to use a little space at the very end of this first part of the book to discuss it.

5.5 Creativity

Some people may wince at the word creativity in a book about scientific method. The educational system often drills us to associate science with thinking styles that are strictly rational. Without reflecting we see poets and artists as creative people, scientists and engineers as rational people, and we assume that these ways of thinking have nothing in common. The truth is that creativity research studies thinking in various fields, not least in science. When Einstein replaced force with spacetime he was not merely rearranging known parts in a normal problem-solving process; he invented a completely new way of seeing gravity. When Newton associated the falling apple with the moon he was not thinking rationally. Neither was Kepler, as we have seen, when he tried to understand the planetary motions. As science aims to generate *new* knowledge and ideas, creativity is an essential part of scientific thinking. This section is too short to even begin to introduce creativity as a subject, but it is so often overlooked in research education that it must be allowed a little space here. We will very briefly look at how creative people think but first we shall discuss how our expertise can be detrimental to creativity. This is an important point for researchers who are often experts in their fields. It can be understood using a theory of learning that is attributed to the psychologist Abraham Maslow. I will call it the four stages of learning and use a simple example to explain it:

Before Newton invented differential calculus, mathematical problems were solved using geometry. Those who have learned geometry in school know how difficult it is. You must train your brain to think in a certain way that initially feels awkward but gradually becomes more natural. Imagine that you are about to learn geometry for the first time. Before you have tried it you have no notion of how difficult it is. This is the first stage of learning where you are *unconsciously incompetent*; you do not yet know that you cannot do it. After a couple of lessons you start to understand the basic ideas of geometry but you realize that you are not able to use it at all. You are now *consciously incompetent*. Despite this insight you do not give up. After a couple of weeks you find that you have begun to understand how to solve various problems in your geometry book but you must think carefully about the rules that you have learned to be able to apply them correctly: you are *consciously competent*. Now, imagine that you have used geometry for many years to solve various mathematical problems. At this point you no longer have to think about the laws of geometry to apply them. Geometry has become second nature and you are *unconsciously competent*, which is the fourth stage of learning. It usually takes several years to reach this highest level of skill because learning how to do things in new ways is simply very hard.

Picture yourself as a geometry expert, used to solving almost any problem with your geometry toolbox, and that someone tries to convince you to switch to infinitesimal calculus instead. Even if they explain the benefits of this alternative method it requires great motivation on your part to "unlearn" your existing expertise and start anew from the beginner's level. You would, after all, learn a different method for something that you already know how to do. Even more motivation would be required to *invent* the new method yourself, as Newton did with infinitesimal calculus. He was highly skilled in geometry but still made the effort to reinvent mathematics to develop a new branch of physics. Where does such motivation come from? You must certainly be very interested in getting to the bottom of the problem that you are studying.

After working in a field for some time we tend to become specialists. We learn to solve most types of problems using the common tools of our trade. The comfort of this expertise often makes us blind to alternative lines of thought and that makes us less prone to thinking creatively. If you only have a hammer, as Maslow said, you tend to see every problem as a nail. It is only natural to prefer to do things in a familiar way rather than in a new and difficult way. As experts, we tend to use the same old ideas over and over.

Quite a few important breakthroughs in theoretical physics were made by people who had not yet reached the age of 25. Newton and Einstein are only two examples out of a good half dozen, and there are many examples outside of physics too. The philosopher Kuhn wrote that "almost always the men who achieve [...] fundamental inventions of a new paradigm have been either very young or very new to the field whose paradigm they change" [8]. It is interesting that people who are capable of true flashes of genius at an early age seldom get flashes of the same magnitude when they get older, despite the fact that their knowledge deepens throughout their lives. Knowledge does not seem to be a limiting factor for creativity. If we do not wish to conclude that the brain suddenly becomes less efficient in connection with our 25th birthday, we may interpret this situation using the four-stage model of learning. Maybe these geniuses were capable of developing truly novel ways of approaching their problems *because* they had not yet developed deep expertise? Without the tunnel vision of a specialist it is easier to be open to a wide range of ideas. Maybe it pays to free oneself from unconscious competence and to deliberately try to be a consciously competent amateur from time to time?

Michalko [9] says that creative thinking has much in common with evolution by natural selection. The basis of evolution is variation, because without variation there is nothing to select from. In a similar manner, creative people are good at generating a wide variety of ideas about a problem before choosing the one to proceed with. He exemplifies this way of thinking with Leonardo da Vinci, who is known to repeatedly have restructured his problems to see them from different angles. He thought that the first approach was too biased towards his usual way of seeing things. With each new perspective he would deepen his understanding of the problem and begin to see its essence. He called this method *saper vedere* – knowing how to see [9]. At first sight this way of thinking may seem wasteful as most of the ideas will never come to any direct use. The point is that, by repeatedly seeking different approaches, we gradually move from our common way of thinking to new ways. Once in a while this process will result in a truly new and useful idea, which makes the whole effort worthwhile.

The Nobel Prize laureate Richard Feynman is another example of this type of thinker. He was once asked how to keep fresh in one's reasoning to avoid stifling the creative process. He replied "All you have to do is, from time to time – in spite of everything, just try to examine a problem in a novel way. You won't 'stifle the creative process' if you remember to think from time to time. Don't you have time to think?" [10]. He clearly did not consider examining a problem with established methods to be thinking at all. It is probably a more common problem than we would wish to admit, that we do not invest enough time in thinking. "Thinking is the hardest work there is", Henry Ford wrote, "which is the probable reason why so few engage in it".

The inevitable question is if creative thinking can be learnt. Fortunately, it seems like it can. In the 1970s, Zuckerman [11] studied Nobel Prize winners in the United States. It certainly requires creativity to make scientific contributions unique enough to be rewarded with such a prestigious prize. Interestingly, she found that they had often been trained by other Nobel laureates. For example, six of Enrico Fermi's students won the prize. Ernest Lawrence and Niels Bohr each had four students who won them. J.J. Thomson and Ernest Rutherford between them trained no less than seventeen Nobel Prize winners. The students testified that the key to their success was the different thinking styles and strategies that their teachers had taught them, rather than the specific knowledge. In some cases their teachers even knew less than them within their specific field of interest [11]. This shows that their teachers obviously could enhance their creative abilities. Regardless of the level we start from, there is probably hope that we all can improve our thinking styles.

The key to becoming more creative is to avoid thinking *reproductively*. This is the ordinary approach when confronted with a problem – thinking of how we have solved similar problems in the past. Creative people think *productively*. Instead of asking how they have been taught to solve a problem, they ask how many unique ways there are to solve it [9]. There is a wealth of books on techniques for creative thinking and this chapter is too short to provide any details. A couple of simple graphical techniques will be introduced in Chapter 10, which deals with the planning phase of an experiment, and visualization is indeed one way to explore problems from different perspectives and to capture their essential features. The notebooks of great scientific thinkers are often scribbled over with sketches and diagrams. Other ways are to rephrase a problem in different ways or to put it into completely different contexts to discover its different aspects.

This short introduction will necessarily be unsatisfactorily superficial but the message is that scientific method would not amount to much without creativity. Unfortunately, this is often overlooked when research methods are discussed and the result is often a false impression of research as a strictly rational activity. In reality, scientists are a bit like jazz musicians and experiments are like the music they play. Despite their deep knowledge and long experience, scientists often do not know exactly how they are going to arrive at an answer. They improvise and create to get there. Research into a problem can be seen as the exploration of a theme. As in music, the journey is more important than the destination. Although the end result of the research may have profound effects on scientific understanding when it has been filed in the annals of science, that wisdom was acquired during the detours that were made along the way.

The magic of a research program begins already when it is no more than a twinkle in the researcher's eye.

5.6 Summary

- Research and development may share techniques and practical tasks but they differ in purpose. Research aims to generate general knowledge, whereas development aims to devise specific systems.
- Research has the following four defining characteristics. Firstly, it is based on an open system of thought. Secondly, researchers examine data critically. Thirdly, researchers generalize and specify the limits on their generalizations. Fourthly, research is a self-corrective process.
- Theories are systems of statements about theoretical concepts that represent aspects of the real world. Good theories are simple (parsimonious) and testable. They are also general and useful representations of reality.

- As science aims for new knowledge and ideas, creativity is an essential part of scientific thinking. Creative thinkers avoid thinking reproductively. They try to find several unique solutions to a problem before choosing which one to proceed with.

Further Reading

Those who are interested in scientific realism are recommended to read the first part of Hacking's book [12]. It also contains references to other texts on the subject. There is a wealth of various types of books on creativity. Michalko [9] provides a useful starting point that is more practical than academic.

Answers for Exercises

5.1 Many Ph.D. students in applied fields initially think that they are doing development after filling this diagram out, because they may work with specific systems like engines, drugs, control algorithms or measurement techniques. It is, however, quite possible to do this with an aim to generate general knowledge for others to build on. If your purpose is to generalize, understand and explain, you are doing research. If the purpose is only to get something to work, it is development. In that case, you should probably discuss the contents of your research studies with your academic advisor.

5.2 Left to the reader.

5.3 If the criteria are not fulfilled, you should discuss your research with your academic advisor.

References

1. Mischke, C.R. (1980) *Mathematical Model Building: An Introduction to Engineering*, Iowa State University Press, Ames (IA).
2. Leedy, P.D. and Ormrod, J.E. (2005) *Practical Research: Planning and Design*, 8th edn, Pearson Prentice Hall, Upper Saddle River (NJ).
3. Phillips, E.M. and Pugh, D.S. (1994) *How to Get A PhD: A Handbook For Students and Their Supervisors*, 2nd edn, Open University Press, Buckingham.
4. Molander, B. (1988) *Vetenskapsfilosofi*, 2nd edn, Thales, Stockholm.
5. Feynman, R. (1995) *The Character of Physical Law*, The Modern Library, New York (NY).
6. Dec, J. (1997) *Conceptual Model of DI Diesel Combustion Based On Laser-Sheet Imaging*. SAE Special Publications, **1244**, 223–252.
7. Pickett, L.M., Kook, S., Persson, H., and Andersson, Ö. (2009) Diesel Fuel Jet Lift-Off Stabilization in the Presence of Laser-Induced Plasma Ignition. *Proceedings of the Combustion Institute*, **32 II**, 2793–2800.
8. Kuhn, T. (1962) *The Structure of Scientific Revolutions*, The University of Chicago Press, Chicago (IL).
9. Michalko, M. (2001) *Cracking Creativity: The Secrets of Creative Genius*, Random House, New York (NY).
10. Feynman, R. (2005) *Perfectly Reasonable Deviations From The Beaten Track: The Letters of Richard P. Feynman*, Basic Books, New York (NY).
11. Zuckerman, H. (1977) *Scientific Elite: Nobel Laureates in the United States*, The Free Press, New York (NY).
12. Hacking, I. (1983) *Representing and Intervening*, Cambridge University Press, New York (NY).

Part Two
Interfering with the World

6 Experiment!

It can hardly be said often enough that a well controlled experiment, especially if conducted in the field, is worth more than a thousand observations.

—Staffan Ulfstrand

As mentioned in Chapter 4, there are, in principle, two ways to observe the world. One is to observe without attempting to influence: standing back, watching and waiting for something interesting to happen. The other way is to experiment: actively interfering with the world to create an interesting response. The word experiment comes from the latin *experior*, which means to try or test. In this chapter we are going to look at how experimenters answer research questions by interfering with the world.

6.1 What is an Experiment?

To discuss experimental method we need a definition to set it apart from mere observation. Since the experimenter interferes with the world to obtain useful data, we could say that an experiment involves a *strategic manipulation* of a system to create an *organized response*, in order to answer a *specific question*. These three criteria are a good start, but are they sufficient? When putting some oranges on the scales at the fruit market to find out what we should pay for them, we are making a strategic manipulation of a system (the scales) to create an organized response (the readout), in order to answer a specific question (what should we pay). Most people would probably agree that this is no more an experiment than looking at the speedometer when driving your car. Setting up a system and obtaining information from it by passive observation is not experimentation. When Mr. Green counted his apples in Chapter 2 he was not experimenting. If he had tested different fertilizers to increase the yield of apples, he would have been conducting an experiment.

> **Exercise 6.1:** Before reading further, take a minute to reflect on why the fertilizer test would be an experiment, and just counting the apples is not.

An experiment imposes a treatment on something. It exerts a stimulus on the system under study. As experimenters we are interested in the response to this stimulus, because we hope it will say something about how the system works. If we were interested in how the fruit scales worked and put different objects on them to see how the readout was affected, this would be an experiment – an active interaction with the system under study. If we were only interested in the weight of the oranges, weighing them cannot be considered to be

---

Experiment!: Planning, Implementing and Interpreting, First Edition. Öivind Andersson.
© 2012 John Wiley & Sons, Ltd. Published 2012 by John Wiley & Sons, Ltd.

Figure 6.1
A passive observer investigates a pre-existing state without attempting to influence it.

an experiment. This is because the response we are interested in is not created by us. The oranges would weigh the same whether we weighed them or not.

The crucial difference is this: the passive observer investigates a *pre-existing state* (Figure 6.1). The experimenter *changes the state* and investigates the result of the change (Figure 6.2). Would you, for example, conduct an experiment if you weighed yourself on the bathroom scales every morning? No you would not, because you would be measuring something that was already there. On the other hand, weighing yourself regularly while testing a new diet *is* an experiment, because you are measuring a response to a stimulus that *you* created.

But what is the relevance of this discussion? The topic of this book is experimental method. It would be pointless to discuss this method if it could not be distinguished from other methods. We need some defining characteristics before we proceed. Another important point is that understanding the potential of the experimental method makes it is easier to make good experiments; this aspect is discussed further in the next section. Many Ph.D. students are experimenting intuitively, collecting data without a clear question in mind and trying to extract interesting information from the data afterwards. This is very similar to passive observation. An effective experiment addresses a specific question and requires the development of a clear strategy.

As we concluded in Chapter 4, experimentation is a much more efficient way of obtaining knowledge than passive observation. By creating conditions that are relevant to a specific question we ensure that the information we obtain is relevant. Sometimes, scientists must develop new instruments and apparatus to measure things, but no matter how cleverly a measurement system is conceived or how complex it is, just measuring something does not turn an investigation into an experiment. Experimenters do not settle with just counting or measuring things. If we agree that an experiment must measure a response to a stimulus exerted by the experimenter, this means that there is no fundamental difference between measuring the gravitational constant, counting neutrinos in a neutrino observatory, mapping the cosmic background radiation with a satellite, counting the apples in your garden or measuring the length of a cucumber. They are all forms of passive observation.

It may seem like an unfortunate consequence of our definition that it excludes some classical "experiments" in the history of science, such as the ingenious "Cavendish

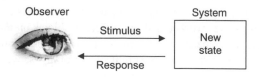

Figure 6.2
An experimenter changes the state and investigates the resulting response.

experiment", which was the first to yield an accurate value of the universal gravitational constant. Cavendish basically devised a very clever balance that measured the gravitational force between lead balls without being disturbed by the earth's gravity. Saying that this is not an experiment is not to belittle the ingenuity and sheer beauty of his setup, nor to belittle its relevance to science. To define what an experiment is, a line must simply be drawn somewhere between experiment and passive observation.

The word "passive" is used here to contrast against the active, interfering nature of experimentation. It may be perceived as a charged word, so it should be stated that passive observation is also a very useful and important method in science. This is an undisputable fact, since there are many examples of great theories that have sprung from passive observation. For example, Edwin Hubble discovered that the degree of redshift observed in the spectra of galaxies increased with the galaxies' distance from the earth. This helped him establish that the universe is expanding. The birds that Charles Darwin collected on the Galapagos Islands during his voyage on HMS Beagle played an important role in his development of the theory of evolution by natural selection. Alfred Wegener noticed that the continents of the earth seem to fit together like a jigsaw puzzle and suggested the hypothesis of continental drift. This was later developed into the modern theory of plate tectonics. All of these important discoveries were made through passive observation by attentive, shrewd observers. In many fields of study it is simply not possible to perform experiments, and in most other cases you must of course first notice something interesting before you are able to investigate it experimentally. Passive observation has contributed tremendously to the growth of scientific knowledge but it is not the topic of this book. The important point is that when we do have a choice between passive observation and active experimentation, we should choose experimentation because it is a more efficient way to learn how things work.

Many students at the beginning of their Ph.D. only have vague ideas about how to conduct experimental research. I once asked a student why he had chosen to manipulate a particular set of variables in his first experiment. He answered "well, you must change something, otherwise it is not an experiment". After graduating he agreed that this statement, though basically true, was a bit naïve. Part of the reason for the confusion may be that school often gives us the false impression that experimentation is mainly about collecting data. During traditional laboratory classes we have used setups that were tested and tuned for us by our teachers, and followed predetermined steps to collect data. These activities have pedagogic value but lack an important aspect of real experiments: the *development of a strategy* to address a research question. Real experiments are preceded by a period of preparation where the idea for the experiment is developed and perfected. We will soon look at some illustrative examples of this, but first we will discuss the importance of how we formulate research questions and how we go about answering them.

6.2 Questions, Answers and Experiments

The reason why it is important to distinguish between experimentation and passive observation is that they provide different kinds of evidence. An observational study reveals *correlations* between variables whereas a well conducted experiment can provide evidence for *causation*. This is because active and structured manipulation of a system can isolate the effect that one variable has on another. In an observational study we do not take control over the variables, which makes it difficult to work out "what does what to what" in the system.

In Chapter 2 we compared the inductive and hypothetico-deductive approaches to research and concluded that they provide different sorts of knowledge. We said that the inductive approach could only give us descriptive theories, whereas an hypothesis could be a basis for explanatory theory. These two approaches have nothing to do with our definition of experimentation. Both experiments and observational studies can test hypotheses but neither of them has to be hypothesis-driven. If we test hypotheses or not is a matter of how we formulate our *research questions.*

By contrast, if we experiment or not has to do with how we go about *answering* the questions. As explained in Figure 6.3, these are two completely independent dimensions of scientific enquiry. The two types of study (observational and experimental) and the two types of research question (non-hypothetical and hypothetical) give us four possible combinations. I have chosen not to use the term inductive in this figure, since philosophical purists may argue that an inductive study is an impossibility, and the crucial point is really if the research question is based on an hypothesis or not. If it is not, the knowledge obtained can only be descriptive. If it is, it is at least possible to obtain explanatory knowledge from the study. Whether we do or not depends on the nature of the hypothesis.

Recall from Chapter 5 that phenomenological theories describe relationships between variables, whereas explanatory theories describe the mechanisms behind phenomena. Hypotheses can be of both these kinds. For instance, an hypothesis stating that a certain nutrient decreases the blood pressure can only result in phenomenological understanding, because it only makes an assertion about the relation between two variables without explaining *why* there is a relationship. If we, on the other hand, state that the nutrient decreases the blood pressure *because* a substance in it interacts in a specific way with the cells in the blood vessels, then testing this hypothesis will give us insight into whether the explanation is true or not. In this way, an hypothesis involving a mechanism provides explanatory knowledge. As previously mentioned, this type of knowledge is the long-term goal of science, though descriptive knowledge can be an important intermediate goal.

It is perhaps easiest to clarify the combinations in Figure 6.3 by some simple examples. Say that you discover, during your travels at the far end of the world, that people in two villages show very different incidences of hay fever. You hear that the villagers of the

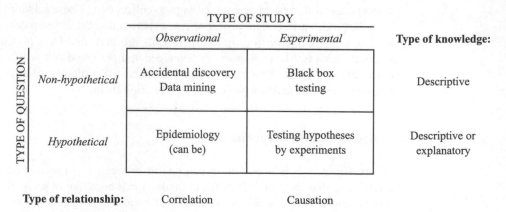

	TYPE OF STUDY		
	Observational	*Experimental*	**Type of knowledge:**
Non-hypothetical	Accidental discovery Data mining	Black box testing	Descriptive
Hypothetical	Epidemiology (can be)	Testing hypotheses by experiments	Descriptive or explanatory

(TYPE OF QUESTION)

Type of relationship: Correlation   Causation

Figure 6.3
The research question and how it is answered (type of study) are two independent dimensions of scientific enquiry. Different types of questions lead to different types of knowledge, and different types of studies give evidence for different types of relationship.

community where hay fever is uncommon regularly eat a specific herb. As this herb is not eaten in the other village, you suspect that it causes the difference. This discovery was not preceded by the formulation of an hypothesis, nor did it result from an experiment. You just made an accidental discovery of a potential relationship between two variables, and thereby find yourself in the upper left box in Figure 6.3. As long as you have not conducted a structured study your case is very weak. It is based on anecdotal evidence for the effect of the herb. The knowledge obtained, if we can call it that, is purely descriptive: a village where people eat the herb has few cases of hay fever. We do not know if one causes the other and much less why it would.

The other example in the upper left box is data mining, which is a more reliable method than accidental discovery. It uses statistical methods to look for relationships between variables in large data sets. Since the method explores a very large number of potential relationships, spurious relationships are sometimes found. These are accidental correlations between unrelated variables. The fact thereby remains that observational studies do not give good evidence for causation and, if there is no hypothesis, they cannot provide explanatory knowledge.

Now, your discovery in the two villages could inspire you to make a larger study. Say that you formulate an hypothesis stating that eating the herb somehow reduces the risk for contracting hay fever. To build a case you start collecting data about a large number of people in an epidemiologic study. Other possibilities must be excluded, such as social, hereditary and environmental factors that also could influence the risk for hay fever. Your data set must, therefore, go far beyond whether people eat the herb and whether they have hay fever. This is now a hypothesis-driven study. Since your hypothesis does not suggest an explanation of why the herb is effective, it can only result in descriptive, phenomenological knowledge. To build an explanatory theory you must formulate an hypothesis that includes a mechanism, which may be possible at a later stage. As you are not actively administering the herb, however, this is still an observational study and you find yourself in the lower left box in Figure 6.3. (Incidentally, it should be noted that not all epidemiological studies are confined to this box.) Observational studies are improved by using larger amounts of data as this decreases the uncertainty in the conclusions. Although this study is much larger and more structured than the first one, it still does not provide good evidence for causation. To show that variation in one variable is the actual cause of a response in another, you must actively manipulate that variable in a structured way.

To demonstrate causation you must, in other words, conduct an experiment. You could, for example, devise a blinded experiment to test the hypothesis that the herb decreases the risk for hay fever. Such a study involves a control group that receives a placebo, whereas the experimental group receives the herb. It is conducted in a standardized fashion so that the substances are taken in the same quantity at the same intervals under the same conditions. The results will now provide good evidence of causation, since the experiment isolates the potential effect of the herb. The control group excludes the possibility that an effect is due to background variables that are outside the experimenter's control. The knowledge obtained is, however, only descriptive. To *explain* the effect the hypothesis must involve a mechanism. You could, for instance, formulate an hypothesis about how a specific substance in the herb interacts with the immune system and test it in a cleverly designed laboratory experiment. In that case your study could both give good evidence for causation and explanatory knowledge. Whichever hypothesis you use, you find yourself in the lower right box of Figure 6.3.

It may seem confusing that a study in the lower left box could be the basis of explanatory theory but not give evidence for causation. There is, however, no contradiction in this. Data can provide *support* for an explanatory hypothesis but never *prove* it. Evidence for causation is stronger than a mere correlation but it is still not proof for causation. There is always a degree of uncertainty in our conclusions. To make the strongest possible case for our conclusions we should strive to be as low and as far to the right in Figure 6.3 as possible.

The only box we have not visited yet is the upper right one. How can an experiment be conducted without an hypothesis? Well, it is fully possible to expose a system to different treatments just to find out what happens. This is probably how Brewster managed to induce birefringence in materials under stress, as mentioned in Chapter 2. We could call this type of study "black box testing", as it is more concerned with how something behaves than how it works. It exposes a system to various treatments to produce reactions, without preconceived ideas about its internal workings. This is actually a common method for testing software. Due to the active manipulation of the system the results can give good evidence that the output is caused by the input, but it says nothing about why.

Now that we have sorted out some fundamentals, it is time to ask ourselves how experimenters, in practice, go about answering research questions. The best way to do this is perhaps to look at a number of real world examples.

6.3   A Gallery of Experiments

The following seven experiments are chosen from various fields and times to demonstrate the generality of the experimental approach. To avoid lengthy explanations of technical terms they are presented in a somewhat popularized form. Many important experiments in the history of science are, when you look closer, long series of experiments that do not lend themselves to brief description. Due to the limited space I have chosen these examples for their relative simplicity. They are often conducted within a limited time span and often involve equipment and concepts that are relatively easy to understand for a layman. Many represent scientific breakthroughs while a couple of them come to more mundane conclusions. The exhibition is chosen for its pedagogical value, not to present an experimental hall of fame.

One problem with research papers in this context is that they seldom discuss how the experimenters were thinking when they developed their ideas. We only get an account of how the experiments eventually were conducted. To amend this situation I have included some experiments described in biographical books, and I have also taken the liberty of including an experiment where I have participated myself.

After touring the exhibition, a number of exercises will help us reflect on how the previous discussion applies to real experiments. It is my hope that this will give the reader a more concrete comprehension of the experimental method than a purely theoretical discussion can provide.

Example 6.1: Secrets of the heart (Physiology)   William Harvey entered the University of Padua in Italy while Galileo was still teaching there. After receiving his M.D. in 1602 he returned to England and began to study the heart and blood vessels. Physiology and anatomy studies in Harvey's day were dominated by the ancient philosophies of Aristotle and the Roman physician Galen. Galen had taught that blood was somehow generated from food in the liver and transported to the heart through the veins. From the heart, arteries transported blood and a kind of "vital air" or "spirit" to the organs of the body, where they were somehow consumed. The relation between these two bloodstreams was unclear. After years of experiments and observations of the heart and blood vessels, both in animals and humans, Harvey published a groundbreaking book on the topic in 1628 [1]. The following is based on that book.

Though every schoolchild today knows that blood circulates through the body, it had not even been imagined in Harvey's day. Some of his contemporaries believed that the pulse and respiration had the same purpose: to ventilate and cool the blood. They did not even understand that the heart caused the pulse. The arteries expanded by themselves, they thought, to take in air through the skin. Others believed that the contracting arteries pumped blood to the heart, and that the heart pumped it back into the arteries when it contracted. It was Harvey who established that the pulse originates from the heart.

His contemporaries also had not realized that the heart pumps blood through the lungs. One reason for this was that the blood vessels become too small in the lungs to be seen by the naked eye. It was known that the right ventricle of the heart was connected to the lungs by the pulmonary artery but it was unclear in what direction the blood was moving through it. As the pulmonary artery branched out into ever-thinner blood vessels in the lungs, they eventually seemed to disappear in the tissue. Harvey realized that the valves of the heart ensured that blood could only move from the right ventricle to the lungs, and not back again. It must thereby pass through the lungs and the pulmonary vein into the heart's left atrium, as outlined in Figure 6.4. If blood could pass through the dense tissue of the liver, and if urine could pass through the kidneys, why should not blood be able to pass through the lungs? This thought made him wonder if the two blood streams were not connected in the body as well. Was the blood in fact circulating through the body?

People had not realized that blood passed from the arteries to the veins in the body because, as in the lungs, the blood vessels in the limbs and organs simply became too small to be seen. Harvey used a simple thought experiment to demonstrate that this was the case. He knew from observation that the left ventricle of the heart contained about two ounces of

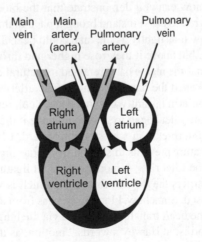

Figure 6.4
Schematic diagram of the heart, its valves and connections to the main blood vessels. Many details are left out. For example, the main vein (*vena cava*) should be two veins; one coming from the upper body and one from the lower.

blood when it was distended. A part of this must be pumped into the main artery when the heart contracted and the valves prevented it from returning into the heart. In half an hour the heart would beat more than a thousand times. Even if only a small part of the blood in the ventricle were expelled at each heartbeat, the total amount pumped from the heart in this time would amount to far more blood than the entire body contains. In a day it would amount to more blood than could reasonably be supplied by eating. The fact that new blood could not be supplied at the same rate as the heart pumped it was evident when thinking about it:

> *Butchers are well aware of the fact and can bear witness to it; for, cutting the throat of an ox and so dividing the vessels of the neck, in less than a quarter of an hour they have all the vessels bloodless – the whole mass of blood has escaped [1].*

But Harvey was not content with just reasoning about the matter. He demonstrated that the heart is pumping blood from the veins into the arterial system by an elegant experiment. He also showed that blood flows from the arteries back into the veins in the limbs of the body by another, equally simple experiment. The first experiment was performed on a live snake and belongs to a category that probably would not be approved by a modern ethics committee. Opening the snake its heart would continue to beat quietly for more than an hour. He had noted that the heart became paler in the systole and got a deeper tint in the diastole, indicating that blood disappeared from the heart when it contracted. Seizing the main vein between his thumb and finger just before it entered the heart, the part between the obstruction and the heart was emptied. The heart became paler, smaller and started to beat more slowly. It clearly pumped blood away from these parts and, due to the constriction, new blood from the vein could not replace it. When he let go of the main vein, all was restored to the previous state. If he instead seized the main artery close to where it left the heart, the part of the artery between his fingers and the heart expanded and got a deep purple tint. The heart pumped blood into this part but it could not escape. If we manage to look beyond our compassion for the poor snake in this experiment, we realize that it demonstrates that the heart pumps blood from the veins into the arteries.

But how can you demonstrate that the blood transported by the arteries to the limbs is transferred to the veins and brought back to the heart? Harvey placed a fillet on a man's arm and drew it as tightly as the man could bear. The arteries beyond the ligature now seized to pulsate, but above it they rose higher and throbbed more violently with each heartbeat. He noted that the man's hand retained its natural color, but gradually became colder. After this he slackened the bandage a little. The arteries above the fillet now shrunk and the whole hand and arm instantly became deeply colored and distended. He also noted that the veins below the fillet began to grow but above it they did not.

The interpretation was straightforward. Blood enters the limbs through the arteries. A tight ligature prevents this blood from entering the arm, which is the reason why the arteries above the fillet rise. A moderately tight ligature allows arterial blood to pass into the arm, but obstructs the flow in the veins, which is why the veins rise below the fillet. The case was closed. Somehow, blood must pass from the arteries of the arm into the veins. From the veins, the heart transfers the blood via the lungs back into the arteries again. The motion of the blood is, in Harvey's words, "motion, as it were, in a circle" [1]. ■

Example 6.2: The cold facts (Engineering)   In January 1986, the space shuttle Challenger suffered a severe accident during launch and the entire crew was lost. Theoretical physicist Richard Feynman was asked to serve on the Presidential Commission that was to investigate the cause of the accident. After careful consideration he accepted the invitation, making a rare exception to his personal rule of never having anything to do with government. The following is based on his own depiction of this work [2].

Working at Caltech he decided to go over to the Jet Propulsion Laboratory (JPL) before leaving for Washington. The engineers there briefed him about the shuttle's technology and everyone was enthusiastic and excited about getting to work with the problem. In Washington, however, Feynman was immediately frustrated by the slow progress. The commission, headed by an attorney, pursued the investigation in large, formal meetings where officials from NASA and other organizations were heard. Some of these were open meetings, broadcast on television. Unfortunately, the persons heard were often not prepared to answer detailed technical questions. Feynman felt that it would be more efficient to talk to engineers that had direct experience and detailed knowledge. He asked to meet with engineers between the formal meetings but was hardly allowed to do so. This would complicate the work of the head of the commission, who had to enter all new information into a record and make it available to all the commissioners. It was a big contrast to the briefing at the JPL, where he had got a large amount of accurate and relevant information very quickly.

A technical detail had caught Feynman's attention during the JPL meeting. During launch, a space shuttle is attached to a large fuel tank with liquid oxygen and hydrogen. On both sides of the tank are smaller rockets for boosting the shuttle. He had found out that these booster rockets were built in sections that resembled pieces of tubing that slid into each other at the joints. The joints were sealed using so-called O-rings; circular rings of rubber that were about $^1/_4$-inch thick. When the rocket became pressurized during launch the sections would bulge between the joints. Rather than being a perfect cylinder, each section would very slightly resemble a wooden barrel and this would open the seals a little, as illustrated in Figure 6.5. To maintain its function the O-ring would have to expand quickly enough to close the gap in the joint. The seals had been known to leak on previous occasions leading to erosion of the O-rings. Such incidents were more common at lower temperatures and it had been particularly cold on the morning of the ill-fated launch. Photos

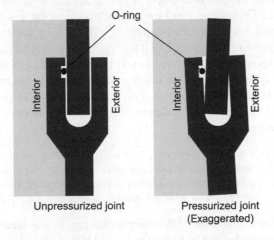

Figure 6.5
Schematic section through a booster rocket joint. As the interior is pressurized during launch, the sections above and below the joint bulge outwards, opening the O-ring seal a little. Adapted from http://history.nasa.gov/rogersrep/v1p60a.htm.

taken during the launch showed black smoke and, later, a flame emanating from a region close to a joint. This made Feynman suspect that the O-rings were somehow a part of the problem.

A fellow commissioner, General Kutyna, also brought the O-ring seal up with Feynman. On the morning of the launch the outside temperature had only been 28–29°F (−2°C). The shuttle had never previously been launched below 53°F (12°C). It was later confirmed that engineers at the company that made the booster rockets were worried about flying the shuttle when the temperature was lower than this. Feynman and Kutyna asked NASA for information about the effect of cold on the O-rings and were promised to get it as soon as possible.

The information eventually came in the form of a thick stack of papers. The first paper said "Professor Feynman of the Presidential Commission wants to know about the effects over time of temperature on the resiliency of the O-rings..." and so on. It was a memorandum to a subordinate. Under that was another, identical memo with the same text from the subordinate to the next subordinate. This continued down to the middle of the stack where there was a paper with some numbers from the last subordinate in the line. Then there was a series of submission papers explaining that the information was sent up through the hierarchy again. There was only one paper with information; the rest were address labels. And the numbers in the middle of the stack were the answer to the wrong question. They showed the effects of temperature over hours, not over the milliseconds that were critical during launch. Feynman realized how inefficient it would be to continue the investigation in this way; hearing people who could not answer crucial technical questions and asking for information through formal channels. "Damn it", he said, "*I* can find out about that rubber *without* having NASA send notes back and forth: I just have to *try* it! All I have to do is get a sample of the rubber" [2]. He got an idea. One of the open meetings would be held the next day and the commissioners were always served ice water during the meetings. He decided to conduct a simple experiment in front of the television cameras.

He went to a hardware store and bought a C-clamp and a pair of pliers, which he brought to the open meeting where a NASA official was to be heard about how the seals were supposed to work. Feynman set the stage for the experiment by asking him about the O-ring: "if the O-ring weren't resilient for a second or two, would that be enough to be a very dangerous situation?" The official confirmed this. When a section cut from a booster rocket's field joint was then passed around the table for the commissioners to study, Feynman used the pliers to take one of the bits of O-ring from the joint. He squeezed it in the C-clamp and put it in his glass of ice water. When the official had come to the appropriate place in his presentation, Feynman pressed the button of his microphone to show the results of his experiment in front of the television cameras. He held the clamp up and loosened it as he spoke:

> *I took this rubber from the model and put it in a clamp in ice water for a while. I discovered that when you undo the clamp, the rubber doesn't spring back. In other words, for more than a few seconds, there is no resilience in this particular material when it is at a temperature of 32 degrees. I believe that has some significance for our problem [2].*

The implication of this simple experiment was clear. After continued investigation it was concluded that an O-ring had failed in one of the booster rockets at lift-off. The hot gas that escaped had impinged on the rocket attachment. This led to separation of the attachment and structural failure of the fuel tank. Aerodynamic forces had then promptly broken the shuttle up and the tragedy had been a fact. ■

Example 6.3: Peas and understanding (Genetics) In the mid-nineteenth century biological thinking was on the move, but fundamental mechanisms remained obscure. Scientists understood that cells came from other cells by some sort of division process. They even speculated that all modern cells originated from the same ancient, primal cell. Fertilization and inheritance, on the other hand, were not at all understood. Cells were known to have a nucleus but its role was diffuse. A crucial contribution would come from Gregor Mendel, an Augustinian friar at the convent of Brno, in the present Czech Republic. He would uncover patterns of inheritance that pointed towards its mechanisms. Many of us think that Mendel is famous for discovering the gene but, as we shall see, he had no notion of genes. This description of his work keeps close to Mawer [3] and Mendel's own paper [4].

Mendel enjoyed spending time in the convent's botanical garden. This interest was to put forth a new shoot on the tree of science: that of genetics. Another important basis of this new growth was his interest in mathematics. The convent had sent him to Vienna University for the formal scientific training required for his parish work as a substitute teacher in the local school. In Vienna, he had studied combinatorics – the mathematics of combinations, patterns and probabilities – which was to become a cornerstone of his work.

Mendel focused his work in the convent garden on the garden pea. This plant produces beautiful, butterfly-like flowers (Figure 6.6). In about 1854 he obtained 34 varieties of pea seeds from local suppliers. Over the following two years he grew and bred them to find suitable characteristics to study. His rare experimental talent is obvious already at this stage: where others worked with "more or less" characteristics like size and qualitative descriptions like "many" or "few", Mendel looked for distinct "either or" characteristics that could be counted quantitatively. With his knowledge of combinatorics he understood that it would be easier to analyze such characters mathematically. He finally settled on seven pairs of clear-cut characteristics. For example, one pair was smooth versus wrinkled peas; another was yellow versus green peas. Besides these characteristics, the list included the color of the flowers and distinct aspects of the shapes of the pods and plants.

The garden pea is naturally self-pollinating. This means that pollen is transferred from the stamens to the stigma within the flower already before the bud opens. It was easy for Mendel to open the immature flower bud and remove the anthers before they were ripe.

Figure 6.6
Pea flowers in the author's garden. © Öivind Andersson.

After preventing self-pollination in this way he could pollinate the mature flowers with pollen of his own choice using a camelhair brush. Doing so, he always crossed the plants both ways: when transferring pollen from one variety to another he always transferred it the other way as well. This reciprocal crossing would empirically confirm that each sex contributes equally to the offspring, which was still debated in Mendel's time.

He began the actual experiments in 1856 by crossing all varieties with each other. Plants that had smooth peas were crossed with plants that had wrinkled peas, plants with yellow peas were crossed with plants that had green peas, and so on for all seven pairs of characteristics. In the next year all seeds from these crossings were planted and grown to maturity. In each case the plants looked exactly like one of the parent plants and, therefore, quite unlike the other.

He chose to call the characteristics that appeared in the offspring dominant and represented them by capital letters. The characteristics that disappeared were called recessive and represented by lower-case letters. A cross between *A* and *a* would thereby produce a hybrid *Aa*, which looked exactly like *A*. He looked after the hybrids and harvested the peas, but did nothing more to them. The peas were dried and kept over winter for planting in the spring.

In the third year, 1858, he got the first generation from the hybrids and noticed something interesting. Although all parent plants had shown only the dominant characteristic, the recessive characteristic now reappeared in some of the new plants. He let all these plants self-pollinate and, again, saved the resulting seeds for growing in the spring.

A peculiar pattern emerged in next year's generation: all the recessive types from the last year produced nothing but recessive offspring – they were *pure*. The dominant types from the last year, on the other hand, produced both characteristics. One third of them produced only the dominant character, but the remaining two thirds produced both dominant and recessive offspring, showing that they were *hybrids*. On average, one out of four hybrid offspring was recessive. This is Mendel's famous 3:1 ratio. Today, this ratio is easily explained by the Punnett square (Figure 6.7), which was introduced by R.C. Punnett in the early 1900s. It shows every possible combination of maternal and paternal genes. In a hybrid cross both parents have both genes and each offspring has equal probability of obtaining any one of the four possible combinations. As the recessive characteristic is only

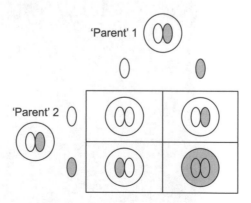

Figure 6.7
Punnett square showing how the dominant (white) and recessive (gray) genes combine. The recessive gene is only expressed when the dominant is absent.

displayed when the dominant gene is absent, the dominant characteristic will be three times as common as the recessive.

Of course, Mendel neither knew of genes nor of the Punnett square. He had to find his own way to explain the ratio. His continued studies followed three lines of investigation. The first had no particular importance for genetics but we must remember that, for Mendel, the experiments were about hybridization, not genetics. He allowed some dominant plants self-pollinate for several generations and confirmed that one third of them produced only dominants, while two thirds produced both dominants and recessives – they were hybrids. He deduced a mathematical rule from this showing that self-pollinating hybrids, if left to their own devices, tend to revert to the pure, parental type over the generations. This is because the hybrids produce both hybrids and pure plants, while the pure only produce pure plants.

Secondly, he crossed plants with combinations of two or more characteristics. For example, he crossed plants having yellow and round peas (*AB*) with plants having green and wrinkled peas (*ab*) and found that each pair of characteristics operated independently of the other pair. Both pairs produced their own 3:1 ratios. Whatever transferred the characteristic from parent to offspring, these things were transferred independently of each other.

Thirdly, he investigated the *composition* of the pollen and egg cells of the hybrids. This was an attempt to explain his observations and here he seems tantalizingly close to suggesting the existence of genes. He reasoned that if a double hybrid (*AaBb*), when allowed to self-pollinate (*AaBb* × *AaBb*), could produce all possible combinations of offspring, it must be due to each plant producing pollen cells and egg cells of all the possible types. He inferred that they produced them in equal numbers; that is 25% *AB*, 25% *Ab*, and so on. If it was a matter of chance which sort of pollen united with which sort of egg cell, the laws of probability implied that each pollen type would unite equally often with each egg type, at least on the average of many cases. He tested this assumption by crossing such a double hybrid with a double recessive plant (*aabb*). This resulted in a 1:1:1:1 ratio of the four characteristics *A*, *a*, *B*, and *b*, confirming his assumption. Today, this type of experiment is known as a "test cross".

It is important to understand that Mendel's letters do not refer to genes. He had no idea about genes, so he could not give them symbols. In his time, biologists had just recently agreed that there were cells and they were still arguing about the meaning of the nucleus. Mendel was merely assigning symbols to what he could see – the appearance of the plants. But when he tried to explain his observations he speculated about the contents of the pollen and egg cells – about "elements" that determine the characters. Groping in the dark he seemed to sense the genes' existence but they remained closed from his view.

Mendel has much in common with Galileo. Just as Galileo with his inclined plane, Mendel used a carefully conceived experiment to uncover a distinct pattern in the workings of the world. As Galileo before him, he rendered the pattern obvious by expressing it in the language of mathematics. In the end he grew thousands of pea plants, so his samples were large enough for him to arrive at the correct ratios and relationships. And, just like Galileo, he laid the foundations of a new science, although the development of genetics would take time. The contemporary scientific understanding had simply not matured enough to give his patterns any meaning.

What Mendel did was, for his time, completely remarkable. To show how remarkable it was, Mawer gives us an example of the "second best" experimenter in the field, Charles Darwin himself. Unknowing of Mendel's experiments, Darwin made his own breeding experiments with two distinct forms of a flower called the snapdragon. He crossed them and found that the first generation only produced one of the forms, whereas some plants in the second generation reverted to the structure of the grandparent. In the words of Mawer:

> *It is pure Mendel. He has even got an approximation to a 3:1 ratio (2.38:1 in fact) with a distinctly small sample. [. . .] So having got this far, what does Darwin do? The answer, I am afraid, is nothing. That is not the measure of Darwin: it is the measure of Mendel [3].*

Mendel's paper on the hybridization of pea plants was published in 1866 in a low-profile journal. It is now considered a key publication in the history of science but the few that read it – among them prominent scientists in the field – simply did not understand its significance. It would take until the year 1900 for the paper to be rediscovered. ■

Example 6.4: The kitchen blender and the stuff of life (Molecular biology)   When Mendel's forgotten paper had been rediscovered in 1900, scientists began to search for the mechanisms of heredity. They believed that the information about how an organism is built was somehow coded chemically in the cells. The sheer complexity of living beings seemed to require a complex molecule to carry the information. For this reason, many thought that proteins were the genetic material. Proteins consist of 20 different amino acids, allowing huge amounts of combinations. DNA had been known since the mid nineteenth century but, since it was built from just four nucleotide bases, it seemed less interesting for genetics. How could the rich music of life be written with just four notes?

In 1952, American biologists Alfred Hershey and Martha Chase made a strong case for DNA being the carrier of genetic information. The following description follows the presentation in their paper [5] from the same year. It is odd to note that Avery, MacLeod and McCarty had concluded already, eight years before them, that DNA carries genetic information [6]. The probable reason that these finding had not caught on in the research community was that they had been published in a journal that was read by microbiologists rather than geneticists [3].

Hershey and Chase were experimenting with viruses that infect bacteria. Such viruses are called bacteriophage, or phage for short. During infection, the phage somehow take over the reproductive functions of the host bacterium and make it produce new viruses. To do this they must clearly transfer genetic material to the host. The problem was to find out if this material was made up of DNA or proteins. Hershey and Chase did this by labeling either the DNA or the protein of the phage with radioactive isotopes. Keeping track of the radioactivity during infection they could see where the DNA and protein went in this process.

They noted that the element phosphorus exists in DNA but not in any of the 20 amino acids that make up proteins. Conversely, sulfur exists in the amino acids cysteine and methionine, but not in DNA. For this reason, they labeled phage with radioactive isotopes of either sulfur or phosphorus. In a first step, bacteria were allowed to grow in media containing either of these isotopes. The radioactive bacteria were then infected with phage and, as a result, new viruses were produced that contained the isotopes that had been incorporated in the bacteria. These labeled viruses were then used in the experiments.

The first experiment investigated whether or not the phage had osmotic membranes. This is a type of membrane that lets water through but is impermeable to salt and other solutes found in living organisms. If the phage were to have such membranes it should be possible to destroy them by what is called osmotic shock. Hershey and Chase began by suspending viruses in a concentrated salt solution. This would create an "osmotic pressure" over the membrane and draw water from the viruses to the solution. This is because water diffusion tends to even out the difference in salt concentration across the membrane. After this they rapidly diluted the solution with distilled water. Reversing the concentration gradient would cause water to flow back into the membranes again, so rapidly that the viruses would burst if they had membranes. And so they did.

What was left after the shock were "ghosts", or viruses incapable of infecting bacteria. This meant that something had happened to their genetic material, whatever it was made of. The ghosts could be separated from the liquid by spinning the solution in a centrifuge, slinging the particles outwards to form a sediment. When they treated sulfur-labeled phage in this way the radioactivity remained with the ghosts. With phosphorus-labeled phage, the radioactivity was found in the solution. This meant that the ghosts were protein capsules and that the shock released DNA from within them to the solution. The ghosts could still attach to bacteria just like active phage but would not cause an infection. The fact that they lost their DNA when they turned into ghosts suggested that DNA might have an active role in their reproduction.

Images from electron microscopes showed that the ghosts were shaped like tadpoles. In modern, higher resolution images they do not look entirely dissimilar to the Apollo Lunar Module. As seen in Figure 6.8, there is a capsule at the top, shaped like an oblong icosahedron (one of the Pythagorean solids). A tail is extending from the capsule. This is the part that attaches to the cell and there is even a landing gear of tail fibers at the base of the tail.

Hershey and Chase started to think that the phage ejected the DNA from the protective capsule when they attached to a bacterium. They decided to test the idea by letting active viruses adsorb to fragments of bacteria. These were obtained from infected bacteria, which produced phage in such quantities that the bacterial membrane eventually would burst and release them. By putting infected bacteria in a special solution that would prevent adsorption of viruses, they made sure that the phage released would not adsorb to the broken membranes. After separating the fragmented bacteria from the solution in a centrifuge, they were put in another solution – one that would promote adsorption – and radioactively labeled phage were added.

It almost worked. After spinning the new suspension in a centrifuge they found most of the radioactive sulfur in the sediment, meaning that the protein capsules had adsorbed to the fragmented bacteria. But when the same procedure was tried with phosphorus-labeled

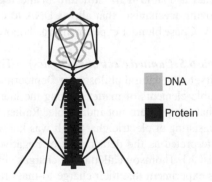

DNA

Protein

Figure 6.8
Appearance of a bacteriophage. The protein capsule at the top contains the DNA. It attaches to bacteria by the base.

phage, about half the phage DNA ended up in the solution. The rest was found in the sediment. This was an indistinct result. It was not possible to tell if the phage released only part of their DNA, or if some of them gave up their entire DNA while others failed to give up any of it. Dissatisfied with the result, Hershey and Chase started to develop what would go to history as the Waring blender experiment.

The hypothesis was still that the phage ejected their DNA into the bacteria during infection. Since they attached to bacteria by their thin tail it ought to be a simple matter to break the phage off the infected bacteria, if the suspension could be subjected to shearing forces that were strong enough. If their hypothesis were correct, this would leave the DNA inside the infected bacteria.

The experiment began by infecting bacteria with phage that were either labeled with sulfur or phosphorus. The unadsorbed phage were removed by centrifugation and the infected cells were suspended in a fresh solution. Running this in a Waring kitchen blender – the kind that is used to mix and smooth out ingredients for food – turned out to strip 75–80% of the sulfur in the phage from the bacteria, which remained intact. The stripped material was found to consist of more or less intact phage membranes. On the other hand, only 21–35% of the phosphorus was released to the solution. About half of this was given up even without the help of the blender. This meant that the bulk of the phage DNA went inside the bacteria. Since the bacteria kept producing new viruses when the parent viruses were stripped from them, the conclusion was clear. The sulfur-containing protein capsules were not needed for the multiplication of phage inside the cell. The main part of the phage DNA entered the cell soon after the phage had adsorbed to the cell.

To confirm that DNA had a genetic function Hershey and Chase investigated the phage progeny: the new viruses that were eventually released from infected cells. They found that less than 1% of the parent sulfur was transferred to the progeny. The parent phosphorus, in contrast, was transferred to the new phage to a substantial extent. This confirmed that DNA entered the cell during infection and took an active role in the production of new phage, whereas the protein capsule remained at the cell surface and had no role in producing progeny. The experiment showed that it was possible to separate phage into genetic and non-genetic parts. It could not rule out that other sulfur-free material than DNA entered the cell but it clearly showed that DNA had an active function in the multiplication of the phage progeny. It thereby strongly suggested that DNA was the genetic material of the viruses.

The year after these results were published, Watson and Crick correctly suggested that DNA had a double helix structure in the legendary journal article that also described the copying mechanism that allows DNA to carry genetic information to new cells. The Hershey–Chase blender experiment was honored with a Nobel Prize in 1969. ■

Example 6.5: Canned electrons (Physics)   The concept of the atom dates back to classical antiquity, to the natural philosopher Democritus who used it to represent the fundamental, indivisible elements of matter. During the late nineteenth century it became increasingly clear that atoms were not indivisible. Rather, they seemed to be complex structures, in part consisting of electric charges. It was known that gases could be ionized; this process was interpreted as the detachment of negative charges from neutral atoms. The British physicist J.J. Thomson called these charges corpuscles. In 1897 he had demonstrated in a famous experiment that their charge-to-mass ratio was always the same. But no one knew if all corpuscles, or electrons as we call them, had the same charge. The question was if electricity was "atomic", that is, if charge existed only in packages of a definite size. How was one to know that electrons were not made from some sort of charged sludge that could exist in any quantity whatever? The only way to answer this question was to make an independent determination of either its charge or mass. Building on the work of others, Chicago physicist Robert Millikan would be the one to provide the answer. The following is based on his own description [7].

Townsend, who was working with J.J. Thomson in Cambridge, was the first to try to determine the electron's charge. He used the fact that water condenses on ions in moist air, forming a cloud. Each water droplet, he assumed, condensed around a single charge. If he could determine the total weight of the cloud and the average weight of the water drops in it, the ratio of these numbers would give the number of droplets. To obtain the charge per droplet he also needed the cloud's total charge, which could be determined using an electrometer.

He determined the mass of the cloud by passing it through drying tubes and determining how much the tubes increased in weight. The average droplet weight could be determined by watching the cloud fall under gravity and using a theoretical relationship called Stokes' law. Apart from the rather optimistic assumption that each droplet contained only one electron, the approach suffered from numerous other uncertainties. For example, it did not account for the influence of convection currents or the evaporation of the drops.

H.A. Wilson was another Cambridge physicist working with J.J. Thomson. He improved Townsend's method by applying an electric field to the negative cloud. He formed clouds by expansion of the moist air between two horizontal brass plates. First he would study how quickly such a cloud sank under gravity alone by observing its upper edge. Then he connected the brass plates to a 2000-volt battery and watched how quickly a cloud would fall under the combined forces of gravity and electricity. This eliminated the need for assuming that each droplet contained only one electron. More heavily charged droplets would fall faster, so watching the top of the cloud he would study only the least charged droplets. A drawback was that he was watching different clouds in the two cases and most of the uncertainties of Townsend's method still remained.

Robert Millikan started by repeating Wilson's experiment and found that the results varied by several tens of percent. It was simply too difficult to make proper observations of the indefinite upper surface of the cloud. He tried doubling the voltage over the plates to increase the accuracy, but was still not satisfied. Presenting his work at a meeting in Chicago, the eminent physicist Ernest Rutherford pointed out that the major difficulty was that the weights of the drops varied with time due to evaporation [8]. This inspired Millikan to take a closer look at the evaporation.

His idea was to hold the charged cloud stationary between the plates by applying an electric field that would exactly balance the force of gravity. This would make it possible to determine the rate of evaporation but it became clear at the first try that the idea did not work as intended. Within a few seconds the cloud had been swept away by the field. He thought that the experiment was ruined but then he noticed that a few drops had exactly the right mass-to-charge ratio to remain suspended between the plates. This serendipitous discovery would allow him to demonstrate the "atomic" character of the electric charge with a method that lacked all the uncertainties of the cloud method. He would call it the "balanced drop method".

The charged water drops could be suspended between the plates for about a minute. If a drop was a bit too heavy at first it would fall slowly under gravity. As it evaporated the downward motion would cease and the drop would become stationary for a considerable time. When the evaporation had continued further it would be drawn to the upper plate. He observed the drops through a short-focus telescope with three equally spaced horizontal cross hairs. The drops were illuminated with an arc lamp and were seen in the telescope as tiny stars against a black background. To avoid excessive evaporation the heat from the lamp was absorbed in three water cells before entering the chamber.

A typical measurement would consist of ionizing the gas in the can containing the plates. This was done using a sample of radium. He then created a cloud by sudden expansion of the volume in the can. He would wait for a droplet to become stationary in the vicinity of the upper cross hair and turn off the voltage. Timing its fall through the cross hairs under gravity allowed him to determine its weight. After that he would turn on the voltage and time its motion up through the cross hairs again. Overall, this method produced much more consistent measurements of the elementary charge than the cloud method. It was even possible to see changes in a drop's movement as individual electrons "hopped on" it during an observation. The atomic property of the charge was plainly visible. The measured charge was always a multiple of one and the same number, which had to be the value of the electron's charge.

Despite this significant improvement of the experiment, there were still uncertainties. The most serious of them was the gradual evaporation of the drops, which also limited the time available for observation. To avoid this problem, Millikan exchanged the water with watch oil in the next upgrade of his experiment. This oil was, after all, specially made not to evaporate. The oil drops were produced and introduced into the chamber using a commercial perfume atomizer. He now realized that he did not have to ionize the gas since the drops became charged by friction in the atomizer's nozzle. Apart from removing an uncertainty, this upgrade also showed that the elementary charge was the same regardless of whether the electrons came from ionization of the air or whether they were picked up in the nozzle of the atomizer. The new method also improved the accuracy by making it possible to study an individual drop for several hours. These observations often kept Millikan working late in the laboratory after a long day of teaching and his social life suffered. On one occasion when he and his wife expected guests for dinner, he was only halfway through an observation of a droplet at six o'clock. He had to call home and ask his wife to go ahead with dinner without him. She explained to the guests that "he had watched an ion for an hour and a half and had to finish the job". One of the guests later said that she thought it was scandalous that the university treated the young professors so cruelly. She had been astonished to hear that Millikan had missed dinner because he had "washed and ironed for an hour and a half and had to finish the job" [7].

Now, is this an experiment or just a measurement? It is true that Millikan used his device to measure the charge of the electron. But it is also true that his setup required an active manipulation of charged oil drops, and in the *structure* of the response he found something more interesting than a numerical value. It revealed something deep about the makeup of the universe. With this low-budget tabletop experiment he was able to demonstrate the atomic property of electrons. This property still holds – no one has split an electron. As far as we know, the electron is one of the fundamental building blocks of this world. One of the beauties of this experiment is that it is simple enough for a high-school laboratory class. Figure 6.9 shows a commercial classroom version, where the metal housing of Millikan's can is exchanged with a transparent cover. The experiment earned Millikan the Nobel Prize in 1923.

But there is something more to learn from Millikan's story. In Chapter 4 we discussed how the experimenter is an integral part of the experiment, and how the development and repeated performance of an experiment can bring out subtle details in its workings. In Millikan's first major paper on the balanced drop method he made no secret of this. He even ranked

Figure 6.9
A modern classroom version of Millikan's setup. His metal can is exchanged with a transparent case through which the two metal plates are seen. Charged oil drops are introduced via the glass nozzle. © Öivind Andersson.

all of his 38 observations from zero to three stars, where the best observations had been made under what he considered to be perfect conditions [8]. He had even excluded some observations and this has been the subject of some controversy. Some have even accused Millikan of fraud. An important point to keep in mind is that his value of the electron's charge would not have been much affected if these observations had been included, but the precision had been poorer. The question is if scientists, who pride themselves upon their objectivity, should allow themselves to exclude data in this way. Looking through Millikan's notebooks, there were good reasons for omitting the observations. There are notes saying "battery voltages have dropped", "manometer is air-locked", "convection interferes", and similar things. Having developed the experiment over time, he had simply come to know what was needed to make a good observation [8]. But if observations are left out for good reasons, why not state that clearly when publishing the results? After all, also scientists are capable of fooling themselves. This is clear when looking at the determinations of the electron's charge that followed on Millikan's. Feynman writes:

> [Millikan] got an answer which we now know not to be quite right. It's a little bit off because he had the incorrect value for the viscosity of air. It's interesting to look at the history of measurements of the charge of an electron, after Millikan. If you plot them as a function of time, you find that one is a little bit bigger than Millikan's, and the next one's a little bit bigger than that, and the next one's a little bit bigger than that, until finally they settle down to a number which is higher.
>
> Why didn't they discover the new number was higher right away? It's a thing that scientists are ashamed of – this history – because it's apparent that people did things like this: When they got a number that was too high above Millikan's, they thought something must be wrong – and they would look for and find a reason why something might be wrong. When they got a number close to Millikan's value they didn't look so hard. And so they eliminated the numbers that were too far off [...] [9] ∎

Example 6.6: Out of the backwash (Engineering) When driving a diesel vehicle, each cylinder of the engine is home to cyclic combustion events that normally take place about 10–30 times every second. Each cycle begins by taking air into the cylinder and compressing it with the piston to a temperature of 900 K or more. Fuel is then injected into the cylinder through a number of small holes in the injector tip. It quickly evaporates in the hot environment and forms jets of fuel vapor that mixes with the hot air until it reaches the auto-ignition temperature. What follows is a chaotic combustion process that both produces the work that propels the vehicle and, unfortunately, a number of chemical byproducts. Some of these are harmful to our health and environment and, for this reason, much engine research focuses on the processes that produce these emissions.

In diesel engines, understanding the processes in the fuel jets is central to understanding the combustion and formation of emissions. The jets are propagating radially into the combustion chamber from the fuel injector, which is placed at the top center of the cylinder. The amount of air mixed into these burning jets determines, among other things, the amount of soot that is formed during combustion (Example 5.1). The mixing process is relatively stable during most of the injection event but, as the injector closes, air entrainment rates increase drastically. After the end of injection a region of increased air entrainment travels from the injector along the jet at twice the original jet speed – a phenomenon called an entrainment wave [10]. This wave may create fuel pockets that contain too much air to burn completely. To an engine researcher this raises the question whether such "overlean" mixtures could be a significant source of unburned hydrocarbons (UHC), which is one of the legislated emissions from engines.

The experiment described here explores a method for avoiding this potential source of UHC in the exhausts from diesel vehicles. As I have been involved in the development of this experiment I should state clearly that it is not a groundbreaking experiment in any sense. I chose it simply because it has two features that are useful for this chapter. Firstly, it is a confined study that produces a distinct result. Secondly, I have first-hand knowledge about how the experiment was conceived and developed. The results were published in 2011 [11].

The story began on a bus trip. A Ph.D. student, Clément Chartier, was planning a research visit to an overseas laboratory and we were discussing ideas for experiments that could be of common interest to him and his host, Mark Musculus. As Musculus had done pioneering research into entrainment waves the discussion soon converged on this topic. Could their potential effect on UHC be mitigated?

A problem with overleaning is that it is a one-way process: once fuel and air have mixed it is very difficult to "unmix" them. On the other hand, if fuel is added to overlean mixtures they become more fuel rich. In other words, if the entrainment wave leaves overlean mixtures in its wake, a small "post injection" (of fuel) after the end of the main fuel injection might render these mixtures combustible. The post injection would of course produce its own entrainment wave, but the change in momentum when the fuel flow is switched off is an important part of how the wave is formed. If the fuel mass and momentum in the post injection were kept small enough, maybe the backwash of air could be kept at a minimum? We agreed that the experiment was interesting because, if successful, it would not only demonstrate a method for mitigating the effect of the wave – it would also allow us to estimate the *importance* of the wave. If we could measure a decrease in UHC emissions due to the post injection, something could be said about how much the wave contributed to them.

Discussing the idea with Musculus we realized that another source of UHC could potentially overthrow the whole idea. After the injector closes, a small amount of fuel often dribbles slowly from its holes and results in fuel-rich mixtures that contain too *little* air to burn completely. An extra injection would result in an extra opportunity for dribble, which could offset the potential positive effect of the post injection. On the other hand, if the post injection were to decrease the UHC *despite* the extra dribble, this would strengthen our conclusions about the wave.

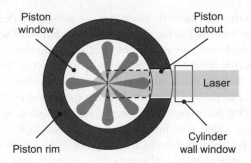

Figure 6.10
Schematic bottom view of the combustion chamber of an optical diesel engine.

The experiment was to be performed in a so-called optical engine. In such engines, parts of the combustion chamber are exchanged with transparent pieces, making it possible to study processes in the cylinder using optical techniques. Figure 6.10 shows a schematic bottom view of the combustion chamber of an optical diesel engine. The combustion chamber is essentially a cavity in the piston. The whole cavity is visible through a window in the piston, which corresponds to the round white area. Eight fuel jets are seen through the window, emanating radially from the centrally placed injector. Laser radiation is entering the combustion chamber from a cylinder wall window and a cutout in the piston. The dashed rectangle indicates the laser-illuminated area that is imaged on a fast camera.

Our first idea was to measure the fuel concentration in one of the jets directly. Illuminating the fuel with a laser we would obtain a fluorescence signal that could, potentially, be quantified. Since commercial diesel fuels contain a wide range of components, this technique would produce a mix of signals that would be difficult to interpret. To avoid this problem, a non-fluorescent model fuel could be used along with a suitable fluorescent fuel tracer. Such measurements are, however, still riddled with uncertainties and assumptions. After some discussion between the research groups the fuel tracer idea was abandoned. A more straightforward laser technique had been developed at the host laboratory. It would only provide indirect information about the fuel concentration but, on the other hand, it was already developed to maturity and could separate between UHC that resulted from lean and rich mixtures. It was thereby capable of distinguishing between rich UHC from injector dribble and lean UHC from the entrainment wave.

The experiments began with a preparatory test, made as a precaution against an effect that could potentially muddle the analysis. It is well known that UHC emissions decrease at higher engine loads. This is because temperatures increase in the cylinder when more fuel is injected and burned, enhancing the oxidation of UHC. The post injection would of course increase the amount of fuel burned. If it were to decrease the UHC, how could one know if it was due to the desired enrichment of lean mixtures, or to unintentionally increased temperatures? To avoid this ambiguity a load sweep was made on the optical engine. The UHC emissions were measured as function of engine load by gradually increasing the amount of fuel injected in a single injection. In later tests with a post injection, the resulting UHC emissions could be compared to this reference curve. If they fell below the curve at the load obtained with the post injection, the reduction would not be an effect of increased load.

An interesting effect was seen when adding a small post injection to the main injection. If the extra fuel was injected shortly after the main injection the UHC emissions fell below the reference curve by about 20% – a substantial decrease. If the extra injection was delayed a bit, on the other hand, the emissions coincided with the reference curve. This meant that the post injection was either effective or ineffective for UHC reduction depending on the injection timing. But was this only an effect of the timing? A special instrument showed that the post injection contained a much smaller quantity of fuel at the early timing than at the later timing. This difference in quantity was likely due to hydraulic effects inside the injector. So, a large post injection was ineffective for reducing the UHC emissions, whereas a very small post injection was effective. This was in line with the original hypothesis on the bus trip: less fuel flow gives less momentum in the spray, resulting in a weaker entrainment wave. In other words, a small post injection seemed to enrich the overlean mixtures from the main injection enough to render them combustible, but without causing significant overleaning itself.

We should remember that, so far, the effect of the post injection had only been seen as a decrease in the exhaust gas UHC concentration. The mechanism behind this decrease had only been hypothesized. As a first step towards finding out, the combustion process inside the optical engine was imaged using a high-speed video camera. From these recordings the effect of the post-injection became evident. With a single injection the flame remained close to the periphery of the combustion chamber. This indirectly showed that any lean mixtures remaining stagnant closer to the injector simply did not burn. By adding the small post injection, however, the flame rapidly crept from the periphery towards the injector. These results were very promising, but to confirm that there were in fact lean mixtures close to the injector, laser spectroscopic measurements had to be made.

The laser technique was now set up in the laboratory. One laser and camera were used to detect radicals of OH, showing the locations of the flame reactions. Another laser and camera were used to detect the UHC. With an extra spectrometer it was possible to see if these originated from rich or lean mixtures. Appropriate optical filters were used to separate the signal from background radiation from the flame, and the two lasers and cameras were triggered so as to obtain simultaneous images of the OH and UHC distributions in one of the fuel jets. The imaged area of the cylinder is indicated in Figure 6.10. Looking at the result in Figure 6.11 we see that the single injection (left-hand image) leaves UHC in its wake. The image was acquired after the injector had closed and the remnant UHC in the jets are seen as dark grey spokes pointing from the injector at the left towards the combustion chamber wall at the right. Analyzing the spectral content of the UHC signal it

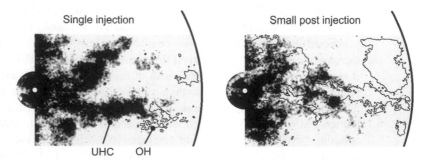

Figure 6.11
Images of the UHC (gray) and OH (solid contours) distributions in the cylinder. The left-hand image shows the distributions after a single injection, the right-hand image after a small post injection.

was found to originate from lean mixtures. The solid contours at the wall show islands of OH, indicating flame reactions in this region. The right-hand image corresponds to the case with a small post injection. It was acquired after the end of the post injection, at exactly the same point of the cycle as the left-hand image. Here, much less UHC is seen in the image, and the flame reactions extend all the way back to the injector.

It was clear that a small post injection had the hypothesized effect of enriching the overlean mixture from the main injection while not producing any significant amounts of overlean mixture itself. But why was the larger post injection ineffective? A closer investigation indicated that the large post injection had sufficient momentum to create a substantial entrainment wave itself, together with the overlean mixtures that came with it. The large post injection probably cleaned up after the main injection but to no avail: it restored the mess that it had swept away. ■

Example 6.7: Sex and the savannah (Ecology)   It is fascinating to reflect that, following your genealogy backwards in time, a long chain of ancestors connects you to the origin of life. Each and every one of them was successful at surviving long enough to reproduce – those who were not successful simply did not become ancestors. In our relative comfort it is easy to forget that their lives were, for the most part, very harsh. They had to compete for food, living space and mates, often under the threat of predators.

Nature's brutal order gradually caused our ancestors to become more fit to their environments. This process is called evolution by natural selection and was discovered by Darwin and, independently, by Wallace. The idea is that certain inheritable traits may either increase the number or the fitness of an organism's offspring, where fitness is a measure of the ability to survive and reproduce. Such inheritable traits make it easier to become an ancestor and, for this very reason, they will tend to become more common in a species over time. Likewise, unfavorable traits will become less common, since they result in less successful ancestors.

Traits that improve an organism's chances of becoming an ancestor include effective camouflage, agility or other properties that increase the ability to escape predators. They may also consist of sensory abilities or physical features that increase the ability to find food. We intuitively understand why such traits have evolved but sometimes Nature produces traits that are simply handicapping. The peacock's tail is just one example of such a feature. It both makes its owner more conspicuous to predators and less able to outrun them. At first sight, the existence of such traits seems to contradict Darwin's theory of natural selection. How could they have evolved?

Darwin finally proposed that the answer was a variant of natural selection that he called sexual selection. It takes place when males combat each other in competition for females and also when females choose a mate that they prefer. In the latter case males compete to be chosen and females choose by looking to resources and other attributes that can be correlated with the quality of the males' genes [12].

This is not to be taken literally of course, since animals generally know very little about genes. The idea is rather that certain mating behaviors in females are expected to give their offspring better chances of survival. If they tend to go for males that have extravagant ornaments, like the peacock's tail, they are more likely to choose a mate with good genes. This is because a male with an impressive ornament probably has a number of favorable properties. He must have been good at finding nourishment to build the ornament in the first place. He must also be resistant to parasites and infections to maintain it. Finally, the more handicapping the ornament is, the more likely it is to signal physical strength and genetic quality. Surviving in the wild with a handicap is, after all, quite a feat. (It should be said that also females often compete over mates, but usually less conspicuously than males.)

So, sexual selection sounds like an interesting theory, but is there any way to test it experimentally? If anyone thought that ecologists limit themselves to passively observing animals from a distance, the following trendsetting experiment will prove them wrong. Swedish biologist Malte Andersson chose to study a peculiar African bird called the long-tailed widowbird [13]. The male and female of this species are normally brownish and very similar to each other, looking a bit like ordinary sparrows. During the mating season, however, the males turn black and grow a tail of about 50 centimeters. Figure 6.12 shows a male widowbird during the courtship display. Being much longer than the bird itself, the tail impedes flight and is severely handicapping. It clearly must have some purpose for mating, but is it to provoke other males or to attract females?

If this ornament evolved through sexual selection, the females should prefer males with larger tails than normal. If they did not, genes for long tails would not have become so common in the population. Andersson presumed that this preference should show itself if females were allowed to choose among males with different tail lengths. He located 36 male widowbirds with territories in the open grasslands of the Kinangop plateau in Kenya. They were carefully sorted into groups of four that contained males with similar tail length and territorial quality. He then manipulated the tail lengths of the birds in each group. Firstly, the tail of one bird was cut to about 14 cm and the feathers were glued to another bird. To make the manipulation of these two birds more similar the outer three centimeters of the tail feathers were glued back on the shortened male. This resulted in extremely long- and short-tailed birds. The remaining two birds in the group were used as controls. One was not treated at all. The other one had its tail cut and glued back on again. This was to see if the cut and glue procedure had any "psychological" or aerodynamic effects that might alter the birds' behavior or flight.

The use of controls is a precaution made to ensure that the results are due to the variables manipulated in the experiment and not to background variables that are outside the experimenter's control. Apart from not being manipulated, the control group should be identical to the experimental group. For this reason, Andersson matched the groups

Figure 6.12
Male widowbird during the courtship display. © Malte Andersson.

according to initial tail length and territorial quality, but kept the tail length of the control group fixed throughout the experiment. As a further precaution, he counted the flight displays and territorial disputes of each male before and after capturing them, to ensure that he did not change their behavior. Another elegant feature of this experiment is that each male was compared to himself. Comparing the mating success before and after treatment for each bird reduces the risk that the analysis is influenced by other variation between the individuals (this will become clearer when we introduce the paired *t*-test in Chapter 8).

The results of the experiment are shown in Figure 6.13. The mating success was measured by the number of active nests in the male's territory. Before the treatment there were only small differences between the shortened, elongated and control groups. After treatment the mating success changed, just as the female choice hypothesis predicted. The mating success was lowest for the shortened males and highest for the elongated ones. This could, in principle, be due to the females not recognizing the shortened males as birds of their own species. On the other hand, the control and shortened groups had rather similar success, which they would not have if shortening had destroyed species-specific features. Andersson also verified that the change was not an effect of variations in activity or territory size. The shortened males were in fact slightly more active, taking off to protect their territory more often and making more frequent flight displays for females. The differences in mating success were apparently due to females finding longer tails more attractive: "The results presented here support Darwin's hypothesis that certain male ornaments are favored by female mate choice, and probably evolved through it" [13].

This means that there is no contradiction between the principles of natural selection and the evolution of features that make it more difficult for a male to survive in the wild. Andersson's innovative experiment made it possible to determine the role of the male widowbird's bizarre tail with greater certainty than an ever so large number of observations. It clearly demonstrates the fact that the experimental method is both more accurate and more economic than passive observation. ■

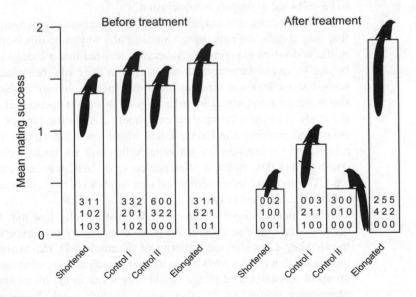

Figure 6.13
Mating success before and after tail treatment. Control group I had the tails cut and restored, whereas control group II was not treated at all. (Adopted from Nature Publishing Group © 1982.)

6.4   Reflections on the Exhibition

Recapitulating our definition, an experiment changes the state of a system to produce an organized response, in purpose to answer a specific question. We added that the response of interest must arise from the treatment imposed by the experimenter. If these criteria are not fulfilled it is an observational study, where measurements are made without attempting to influence the response. One reason why observational studies are sometimes mistaken for experiments is that building and operating measurement devices often require a great deal of practical manipulation, though measurements in themselves are only a way of observation. Looking at the experiments in our gallery we see that they all change the state of a system in order to study the resulting response. For example, Harvey obstructed blood flows in various ways to observe how the vessels were emptied or filled, and Feynman put an O-ring in cold water to see if its resiliency was affected.

> **Exercise 6.2:** As all seven examples in the gallery are experiments they should provide good evidence for causation. For each experiment, give at least one example of a causal relationship revealed by the experiment.

These experiments are captivating in that they use rather simple means to interact with things that often cannot be seen or touched: the minute capillaries that transfer blood from the arteries to the veins, the DNA of viruses, subatomic particles like electrons, not to mention female birds' taste in males. Mendel even studied the effects of something that he did not know existed – genes. It may not always be possible to replace an observational study with an experimental one, but it is equally true that many experiments reveal things that would simply be impossible to find out in observational studies. The only example from the gallery where I can see a possibility for this is the widowbird study in Example 6.7 – but just think of the immense amounts of observational data that would have to be collected to support its conclusion.

We have seen that even experimenters must take precautions against background variables that may muddle their analyses. One common way to do this is to use a control group, as in the widowbird experiment. Andersson realized that a change in mating success could be due to several factors, such as changes in the birds' behavior, the females' ability to recognize the males as birds of their own species, and so on. He included control birds that were not manipulated in order to rule out such influences. If a control group is used, the study is called a "controlled experiment". A similar type of precaution was used in the engine experiment in Example 6.6, where it was known that the UHC emissions were affected by the engine load – a variable that was not directly connected to the research question. For this reason, a reference curve of UHC was acquired as function of engine load. The results of the experimental treatments were then compared to this curve to judge if there was an effect or not.

When looking closer at the examples, we actually find that many of them have no control or reference sets. Does this mean that they are poor experiments? For example, in the Hershey–Chase blender experiment (Example 6.4), radioactively labeled phage were manipulated in various ways to see where the radioactivity went during infection. A control group of unmanipulated phage would only result in the trivial statement that the labeled phage were radioactive. In other words, controls would be superfluous since a control group would not be able to show an effect. The same is true of Millikan's experiment in

Example 6.5, where the studied effect could not have been produced without manipulating the electric field. There was no need for controls in this experiment but, as we have seen, other types of precaution were needed to interpret the results correctly. To mention one example, he eventually used oil drops instead of water to eliminate the problem with evaporation. Not all experiments need a control group but all experimenters need to be vigilant against the effects of background factors.

> **Exercise 6.3:** Determine, for each example experiment, which experimental precautions were taken against background factors.

Finally, in some of the experiments there are random elements that may somewhat dim the effects. In such cases it is important to use sufficient sample sizes to discern the effects and to use appropriate mathematical tools to make the effects visible. We have seen examples of this both in Mendel's experiments and in the widowbird study. It is especially common in biology that the analysis of experiments relies on appropriate use of statistics, but statistics knowledge is important in all data analysis. For this reason, we are going to spend the next few chapters looking at statistical techniques.

Before that, we will round this chapter off with a number of exercises to reflect further on our tour of the gallery:

> **Exercise 6.4:** Look at the examples and identify the research questions addressed by the experiments. If the experiment is hypothesis-driven, state the hypothesis and try to explain how the experimenters arrived at it.

> **Exercise 6.5:** Starting from the research questions identified in Exercise 6.4, explain the ways in which the experiments answer the questions. It is the coherence between the experimental design and the research question that is interesting here.

> **Exercise 6.6:** For each example, explain if the knowledge obtained is explanatory or descriptive and motivate your explanations.

> **Exercise 6.7:** Now that you have trained your eye on the examples, find one or several published research papers that you find interesting. Ideally, they should be relatively concise. They do not have to be papers from your own field – you may get interesting new ideas and perspectives by reading about other subjects! After reading the papers, identify the fundamental research questions and hypotheses that they are based on. Explain, in your own words, the methods used to address the research question. Determine if the investigation is an experiment and, if so, determine if appropriate controls or other precautions were used. Is the resulting knowledge explanatory? Could the experiment have been designed in a better way? It is useful to carry this exercise out in a group with a following discussion.

> **Exercise 6.8:** Now, approach a research problem that is relevant to your own project. Formulate a research question that needs to be answered and then devise an experiment to answer it. To obtain good evidence for causation, attend to potential background variables by precautions such as the use of controls. Good luck!

6.5   Summary

- An experiment changes the state of a system to produce an organized response, in purpose to answer a specific question. The response of interest must arise from the treatment imposed by the experimenter. If not it is an observational study.
- An observational study investigates pre-existing states without attempting to affect the response.
- An observational study reveals correlations between variables whereas a well-conducted experiment can provide evidence for causation.
- Studies based on non-hypothetical research questions can only provide descriptive knowledge. Hypotheses can provide a basis for explanatory knowledge. The prerequisite for this is that the hypothesis proposes a mechanism for the observed phenomenon and not just a phenomenological relationship.
- It is sometimes possible that the response results from variations in variables other than those manipulated in the experiment, so-called background variables. Precautions are needed to verify that such variables are not causing the observed effects. In such cases it is common to use control groups or other reference data.

Further Reading

Biographical books are often a good source of information about how scientists get and develop ideas. Research papers tend to leave these aspects out. Look for biographies of scientists in your field. Crease [8] gives ten short biographies of experiments in physics. Though some of them do not fulfill our criteria for experiments they are well worth reading.

> ### Answers for Exercises
>
> **6.1** See the discussion in the two paragraphs below the exercise.
> **6.2** To express a causal relationship, a sentence must include both a cause and an effect. Here are some instances from the examples. (6.1) The heart causes the arterial pulse. (6.2) Cold causes the O-ring to loose its resilience. (6.3) Crossing a dominant plant with a recessive (cause) results in a dominant offspring (effect), regardless of if the pollen or egg came from the dominant plant. (6.4) Adsorption of phage to bacteria causes transfer of phage DNA to the interior of the bacteria. (6.5) Applying an electric force, opposed to the force of gravity, to charged oil drops causes them to hover, if it balances the gravitational force. (6.6) A post injection with sufficiently low momentum (cause) enriches the overlean wake of

the main injection enough to render it combustible (effect). (6.7) A longer tail ornament in the male widowbird causes greater mating success.

6.3 In examples 6.1–6.2, no precautions are needed because the effects clearly arise from the treatments. (6.3) Mendel's large samples made it easier for him to see the patterns (there is an element of randomness in how pollen and egg combine). He also crossed female and male plants both ways to make sure that each sex contributed equally. (6.4) No controls necessary, but the sequence of experiments shows a gradual exclusion of error sources, such as adsorption to whole bacteria instead of fragmented ones, investigating the phage progeny before reaching a conclusion, and so on. (6.5) Millikan excludes error sources with each new version of the experiment. He decided to study the evaporation, he switched to individual drops, and he switched from water to oil. (6.6) Reference data were acquired during a load sweep using a single injection. This ensured that the post injection did not decrease UHC by increasing the load. (6.7) Control groups were used. Andersson also studied the birds' behavior before and after treatment, and so on.

6.4 (6.1) Two hypotheses: "the blood circulates" and "its motion originates from the heart". They arose from thought experiments and observation (see the quote about the butcher). (6.2) Hypothesis: "cold causes lack of resilience in the O-rings". He arrived at it by bringing together information from a launch photo and discussions with engineers. (6.3) Apparently not an hypothesis-driven study. Question: "how are characters inherited in a hybrid cross?" Still, an hypothesis is tested by the test cross; "a double hybrid cross (*AaBb* × *AaBb*) produces all combinations because a double hybrid produces pollen and egg cells of all possible types". (6.4) Two parallel hypotheses: "the genetic material consists of proteins/DNA". It is not clear how they occurred but this was probably a current theme in the community at the time. (6.5) Hypothesis: "the electron has a definite charge". There was a race towards finding this out at the time. (6.6) Two hypotheses: "the entrainment wave is a substantial source of UHC" and "overlean mixtures can be enriched by a post injection of sufficiently low momentum". The first occurred directly from previous work by Musculus, the second occurred during discussion of the first. (6.7) Hypothesis: "female choice selects for extreme tail length in the male widowbird".

6.5 Left to the reader. Now that we have identified the research questions, it should be obvious how distinctly the experiments address them.

6.6 To obtain explanatory knowledge the hypothesis must be explanatory: it must involve a mechanism. (6.1) Circulation of blood is a descriptive hypothesis, pumping by the heart is explanatory. (6.2) The hypothesis is descriptive. The greater hypothesis (that lack in resiliency caused the accident by leakage at the rocket joint) is explanatory but more evidence is needed to support it. (6.3) Mostly descriptive knowledge but the test cross tests an explanatory hypothesis. (6.4) Descriptive hypotheses. (6.5) Descriptive hypothesis. (6.6) The first hypothesis is descriptive, the second is explanatory as it uses the momentum of the post injection to explain the results. (6.7) Explanatory hypothesis that tests the mechanism of sexual selection by female choice.

6.7 Left to the reader.

6.8 Left to the reader.

References

1. Harvey, W. (1993) *On the Motion of the Heart and Blood in Animals*, Prometheus Books, Amherst (NY).
2. Feynman, R.P. (1988) *What Do You Care What Other People Think?* W.W. Norton, New York.
3. Mawer, S. (2006) *Gregor Mendel: Planting the Seeds of Genetics*, Abrams, New York.
4. Mendel, G. (1866) Versuche über Pflanzenhybriden. *Verhandlungen des Naturforschenden Vereines in Brünn*, **4**, 3–47.
5. Hershey, A.D. and Chase, M. (1952) Independent Functions of Viral Protein and Nucleic Acid in Growth of Bacteriophage. *Journal of General Physiology*, **36**(1), 39–56.
6. Avery, O.T., MacLeod, C.M., and McCarty, M. (1944) Studies on the Chemical Nature of the Substance Inducing Transformation of Pneumococcal Types: Induction of Transformation by a Desoxyribonucleic Acid Fraction Isolated From Pneumococcus Type III. *Journal of Experimental Medicine*, **79**(2), 137–158.
7. Millikan, R.A. (1921) *The Electron: Its Isolation and Measurement and the Determination of Some of its Properties*, The University of Chicago Press, Chicago (IL).
8. Crease, R.P. (2004) *The Prism and the Pendulum*, Random House, New York.
9. Feynman, R.P. (1992) *Surely You're Joking Mr. Feynman*, Vintage, London.
10. Musculus, M.P.B. (2009) Entrainment Waves in Decelerating Transient Turbulent Jets. *Journal of Fluid Mechanics*, **638**, 117–140.
11. Chartier, C., Bobba, M., Musculus, M.P.B., *et al.* (2011) Effects of Post-Injection Strategies on Near-Injector Over-Lean Mixtures and Unburned Hydrocarbon Emissions in a Heavy-Duty Optical Diesel Engine. *SAE International Journal of Engines*, **4**(1), 1978–1992.
12. Andersson, M. (1994) *Sexual Selection*, Princeton University Press, Princeton.
13. Andersson, M. (1982) Female Choice Selects for Extreme Tail Length in a Widowbird. *Nature*, **299**, 818–820.

7   Basic Statistics

There is nothing permanent except change.

—Heraclitus

Knowledge in statistics is crucial when analyzing data. Most students in the natural and engineering sciences therefore take a course that covers descriptive statistics and some more advanced topics, such as confidence intervals, hypothesis tests and so on. Since these concepts are rarely used in courses other than statistics, students tend to forget about them. As a result, many Ph.D. students never venture beyond using purely descriptive statistical methods such as calculating means and standard deviations of data sets. My personal experience is that statistics courses are often taught outside of a practical context. Instead of teaching how to use statistics to solve real problems, the courses tend to focus on statistical theory and textbook problems. This could explain why knowledge often seems to decay so rapidly after the statistics exam. The best way to forget theoretical knowledge is, after all, to never apply it.

The purpose of this chapter is to lay a foundation for the next few chapters. Statistical concepts are introduced that will be important when analyzing experimental data. Since the readers probably have varying degrees of mathematical training I have tried to use less mathematics than many books on statistics do. Many readers will be familiar with much of the contents, especially the first few sections. They might feel tempted to skip the entire chapter but there are probably a few topics that are new to most readers. As these concepts are used later in the book, you should at least browse through the chapter.

This and the following chapters contain examples and exercises with calculations. Some of the example data are from real experiments, while some are completely imaginary. The examples should, therefore, be used only to learn about the statistical techniques.

7.1   The Role of Statistics in Data Analysis

Anyone who has tried to choose one cookie from a tray that comes fresh from the oven knows how difficult it can be. Some cookies are bigger than others, some have a nicer color and some have more raisins in them. We are all vaguely aware that this sort of variation exists everywhere in life. The distance we drive on a tank of fuel varies from time to time, we don't arrive at work at the same time every morning, our hair does not look the same every day – everything varies.

In many cases we do not bother to cope with this variation. If the fuel tank is empty we just fill it up, regardless of the distance we have driven. In other cases we need to take the variation into consideration. If we are required to be at work at a certain time, for

Experiment!: Planning, Implementing and Interpreting, First Edition. Öivind Andersson.
© 2012 John Wiley & Sons, Ltd. Published 2012 by John Wiley & Sons, Ltd.

example, we probably get up a little earlier in the morning to allow for some unforeseen delays along the way. If we are analyzing data from an experiment it is often necessary to account for the variation in much greater detail. The variation in such data comes both from the measurement setup and the process that we are studying. It is crucially important to find ways to cope with this variation in order to draw conclusions that are correct. For example, if the experiment is aimed at detecting a subtle effect it is important to know how much of the variation comes from the measurement setup. If that variation is larger than the one we create by our deliberate experimentation, something must be done to increase the quality of the measurements. We obviously want to be sure that the effects we measure can be attributed to the factors that are varied in the experiment, and not to the noise in the measurements.

In statistics, variation is often called error. The word is not to be associated with mistakes. It's just a word for the inevitable, natural variation that occurs in real-world data. In an experiment, any variation that occurs through unknown influences is called experimental error and the measurement error is often only a small part of it. Variation in raw materials, environmental conditions, the sampling or the process under study may be larger components. With statistics we can separate such sources of error from the effects. We can also quantify the error. Statistics could be said to be a means of stating, exactly, how uncertain we are about our conclusions.

Statistics is both a toolbox of practical techniques and a mindset that helps us understand the world. Statistical thinking helps us interpret natural phenomena that involve random processes, such as chemical reactions, quantum mechanical systems, evolution in biology, and turbulence in fluid mechanics. Likewise, the absence of proper statistical thinking can be an obstacle to proper understanding of what goes on in the world. We often see examples of this in the news media. Newspapers have some favorite subjects, one of which is to constantly warn us that we risk contracting a terminal disease. The headlines may tell us that a certain type of cancer is above average in a particular area, often close to a factory, mobile base station or something equally unreassuring. They want us to read between the lines that there is a causal connection between the two. Health risks should, of course, be taken seriously. If many people are exposed to a risk, even a small increase in the incidence of health problems could have a large effect on the public health. But that is no excuse for careless interpretation of data. It is quite normal for a cancer form to be above average in some areas: it is an inevitable consequence of variation that things are not the same everywhere. Let us consider an imaginary example to illustrate this:

Say that one day the news billboards proclaim that the caries rate among the citizens of Caramel is above the national average. The news story points to the local candy factory as the suspected culprit. It allegedly discharges byproducts from sweet manufacture into the ground water. Even if it did, does the story prove that the factory caused this "outbreak" of caries? The risk of caries can be expected to vary with several factors, such as eating habits, oral hygiene, hereditary factors, social factors, and so on. This means that the risk varies between people and that there will be a natural variation in the incidence. As a consequence, the cases will not be evenly distributed across the country. Depending on where we look we may find a frequency of caries that is below or above average. As we will see when we discuss the central limit effect, it is often reasonable to assume that approximately half of the places will be above average. In fact, finding a value exactly on the average is highly unlikely. Before we can say anything about how extraordinary the situation is in Caramel, we must know something about how large the variation is in the country as a whole.

As experimenters, we need statistical thinking to understand that there is a natural variation everywhere we look and what that entails when we interpret our measurement data. We also need a toolbox of statistical methods to analyze data. To understand the basis of those tools it is necessary to grasp the concepts introduced in this chapter.

7.2 Populations and Samples

A research team may want to investigate how a certain diet affects the blood pressure of healthy people. It would, of course, be impractical to try to convince every healthy person in the world to participate in this study. Instead, they choose to work with a group of people. The group should be representative for all healthy people, so they must be careful to choose the people randomly. For example, they should not choose people working in a single profession, practicing the same sport, or having the same age, since people in such groups can be expected to respond more similarly to the diet than the whole group of healthy people. After testing their selected group, the researchers use statistical techniques to infer something about how the whole group of healthy people responds to the diet.

In this example, the whole group of healthy people is called the population and the selected group is called a sample. The population always consists of the complete set of possible observations and, in this case, it includes all people that have ever lived and those who are not yet born. This is because we are interested in treating any possible person with the diet, not only those who live today. Entire populations are most often so large that it is practically impossible to work with them. For practical reasons, we always work with subsets of populations – samples. To be useful to our investigations they must be random samples. This means that the sampling must be made so that every possible observation occurs with equal probability. A non-random sample cannot be assumed to be representative of the population.

Random sampling is a central assumption both when designing and analyzing many experiments. It is by no means self-evident that the assumption is true. If we want to measure the flight speed of a certain species of bird, for example, it may prove quite difficult to obtain a random sample. The flying ability varies between individuals of a flock, depending on age, state of the plumage and so on. If younger birds were to be more active and fly more often, we would get a biased sample if we simply chose to study any birds that happened to be flying when we were watching. We should choose our sample so that every bird in the flock has equal probability of being studied, and this is no easy task.

Example 7.1: Imagine that you want to determine the average living space in the nation. Investigating each and every household would be impractical, so you want to collect a representative, random sample. A colleague proposes that you use the coordinate system on a map, choose coordinates using a random number generator and then choose the households that are closest to those coordinates for your sample. It sounds like a good idea, but after thinking about it you realize that this won't produce a random sample at all. Some households are in densely populated areas, while others are in rural areas. For a city apartment to be chosen, the random coordinate must be very close to its address. A countryside farm, on the other hand, may be chosen if the coordinate falls within a very large area. It is reasonable to assume that houses in the country tend to be larger than apartments in a city. Such a sample will thereby be biased towards larger living spaces, as country houses have a greater probability of being chosen. ■

> **Exercise 7.1:** Propose a method for drawing a truly random sample from the households in a nation.

7.3 Descriptive Statistics

A gas analyzer is calibrated daily by using a reference gas with known concentration. During the calibration procedure, measurements of the reference concentration and the zero point are made repeatedly before calibration factors are determined. Figure 7.1 shows measurements of the reference concentration from two analyzers. With analyzer 1 the values are scattered over a wide range. As a group the values are centered on the reference concentration but they display a large variation around it. With analyzer 2 the situation is reversed. The variation between the measurements is small but the centering is poor. In statistical terms we would describe these analyzers in terms of their accuracies and precisions. We would say that analyzer 1 has good accuracy but poor precision. Analyzer 2 has good precision, but the accuracy is poor. These terms are often used in connection with measurement data, so make sure to memorize that *precision describes variation*, while *accuracy describes centering*. In this section, some measures of variation and centering are presented.

When we have measured something we often want to get a quick impression of the result. Methods for summarizing the essential features of a data set are collectively called descriptive statistics. They include measures that most readers are familiar with, such as the standard deviation and mean of a data set. They also include graphical tools for presenting data visually, such as histograms, box plots and normal probability plots. These tools give an impression of the variation and the centering, or central tendency, of the data. They may also show the distribution of the data.

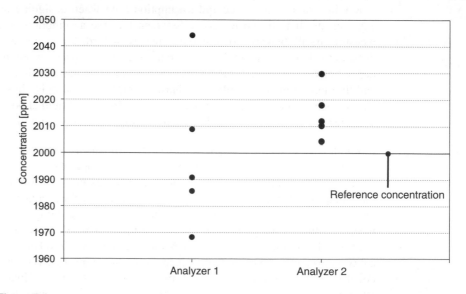

Figure 7.1
Readings from two gas analyzers during calibration. Left: Analyzer 1 has good accuracy and poor precision. Right: Analyzer 2 has poor accuracy and good precision.

The *mean* is perhaps the most common measure of central tendency and is just the average of a set of data. You calculate it by adding all the values in the set and dividing by the number of values in it. The mean of a set of n values is:

$$\bar{y} = \frac{y_1 + y_2 + \cdots + y_n}{n} = \frac{\sum y}{n} \qquad (7.1)$$

where the mean, $\bar{y}$, is read as "y-bar" and the symbol Σ (capital sigma) means that all the y values are to be added together. The *median* is another useful measure of central tendency and is simply the midpoint of a data set. It can be determined by sorting the values from the smallest to the largest. If the number of values in the set is odd there is exactly one value such that half of the values are below it and half are above. This is the median. If the number of values in the set is even it can be divided exactly into two halves. The median is then the value midway between the largest value in the lower half and the smallest value in the upper half, or, in other words, the average of these values. Still another measure of central tendency is the *mode*, which is the most frequently occurring value in a set of data.

Exercise 7.2: Determine the mean, median and mode of the following set of data:

7	2	8	7	7	5	3	15	11

Think about which measure of central tendency you would use if you wanted to find out how well you had done on a written exam compared to your classmates. Comparing your score to the mean score does not necessarily show if you did better than most students. This is because the mean says more about the total score of the group than about the number of students who scored higher than you. To find out if you are in the upper half of your class, for example, the median is a better measure than the mean. In any situation, the relevant choice of a measure of central tendency depends on the question we want to answer.

Quartiles are concepts related to the median. Just as 50% of the data fall below the median, 25% fall below the lower quartile and 25% above the upper quartile. The quartiles and the median thereby divide a data set into four equally large groups. We may also speak in more general terms of percentiles, which are values that a certain percentage of the data fall below. For example, the 80th percentile is the value below which 80% of the data are found. The median and the quartiles are used in a common graphical tool, the so-called box plot, that depicts the distribution of the values in a data set. Figure 7.2 shows a box plot of the data set given in Exercise 7.2. A box represents the mid-50% of the data, which is often called the inter-quartile range as it has the upper and lower quartiles as its upper and lower bounds. Lines extend from the box to the largest and smallest values of the set. Finally, a horizontal line across the box represents the median.

Let us look at measures of variation instead. The most basic one is the *range*, which is simply the difference between the largest and the smallest values in a data set:

$$\Delta y = y_{max} - y_{min} \qquad (7.2)$$

Experiments should generally be designed so that they produce a data range that is larger than the precision of the measurement system. Consider a synthesis experiment in organic chemistry. If we want to measure how the concentration, x, of a certain reactant affects the yield, y, we should vary x sufficiently to produce a range, Δy, that is substantially larger than the precision of the balance that we use to measure the yield. If we do not, the effect

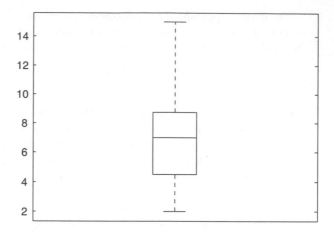

Figure 7.2
A box plot of the data set in Exercise 7.2.

that we are interested in will be indistinguishable from the uncertainty of the measurement. The range is a useful measure because of its simplicity but is too sensitive to extreme values for most purposes.

> **Exercise 7.3:** Figure 7.3 shows dot plots of three data sets, each containing five values. Each dot represents a value. Calculate the mean, median and range for each data set.

A more detailed description of the variation in a data set is given by the *residuals*. These measure how far the data are from the mean. The residual for a value is given by:

$$r_i = y_i - \bar{y} \tag{7.3}$$

where $\bar{y}$ is the mean and y_i is the *i*th value of the data set. Residuals are described graphically in Figure 7.4. The *standard deviation*, which is one of the most common measures of variation, is calculated using the residuals.

Let us assume that we have a sample of n values and want to calculate its standard deviation. Box 7.1 shows all the intermediate steps involved. Firstly, the residuals are calculated, as these describe how far each value deviates from the mean. To make sure

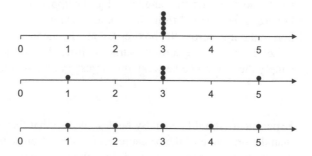

Figure 7.3
Dot plots of three data sets.

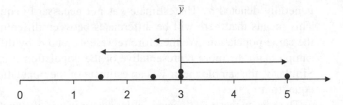

Figure 7.4
The residuals measure the distance from the values to the mean, as indicated by the arrows.

that we are working with positive values the residuals are squared. We then add all the squared residuals up, for every value of i from 1 to n. This sum is called the *sum of squares* and is used so often in statistics that it has its own abbreviation, *SS*. Dividing the sum of squares by $n - 1$ we get a measure of the average deviation from the mean that is called the *variance*. (The fact that we do not divide by the number of observations n may seem confusing and will be discussed shortly.) You may note that the variance does not have the same unit as y. If, for example, y is a distance measured in millimeters (mm) the residuals have the same unit, but the sum of squares and the variance have the unit mm^2. It is often useful to measure the variation in the same units as the mean. This is achieved by taking the square root of the variance and the measure obtained is the *sample standard deviation*,

$$s = \sqrt{\frac{\sum_{i=1}^{n}(y_i - \bar{y})^2}{n - 1}} \tag{7.4}$$

We might ask why it is necessary to first square the residuals and then take the square root. What would happen if we simply summed the residuals? It is easy to convince yourself that the residuals always sum to zero. This will obviously not provide a useful measure of the average deviation from the mean. By squaring the residuals we make sure that all the terms in the sum of squares are positive and that all the values contribute to the standard deviation.

Box 7.1 Intermediate steps in determining the population standard deviation

Residuals: $y_i - \bar{y}$

Residuals squared: $(y_i - \bar{y})^2$

Sum of squares, SS: $\sum_{i=1}^{n}(y_i - \bar{y})^2$

Variance: $\dfrac{\sum_{i=1}^{n}(y_i - \bar{y})^2}{n - 1}$

The reason for calculating s is, of course, that we want to estimate the standard deviation of the *population* from which the sample is drawn. The population standard deviation is generally denoted σ. The estimate s is not necessarily equal to σ due to sampling error. This means that there will be differences between different samples taken from one and the same population. We therefore represent s and σ by different symbols. The larger the sample, and the more representative of the population it is, the better the estimate of σ. Similarly, the sample mean $\bar{y}$ is an estimate of the population mean μ and not necessarily equal to it.

The reason that $n-1$ occurs in the denominator of Equation 7.4 has to do with the fact that we have to calculate $\bar{y}$ to determine s. $n-1$ is called the number of *degrees of freedom* in the sample. This number is related to the number of values in the sample and to the number of population parameters that are estimated from them. It measures the number of independent values in the sample. Consider a sample of $n = 10$ independent observations. Since they are independent, any observation can attain any value regardless of the others. In other words, there are 10 degrees of freedom in this sample. If we now calculate the sample mean the 10 observations are no longer independent. This becomes clear if we reason as follows. If we pick one value from the sample we have no idea what it will be. The same is true if we pick a second and a third value, but when we have picked nine values, we automatically know the tenth. It is given by the first nine values together with the sample mean. This means that there are now only nine independent values, or $n-1$ degrees of freedom left. Since Equation 7.4 requires us to first calculate the sample mean, it dispenses of one degree of freedom and thereby contains $n-1$ in the denominator. The theoretical motivation for using the number of degrees of freedom instead of the sample size in Equation 7.4 goes beyond the scope of this book. We will settle with stating that using n to calculate the standard deviation of a small or moderately sized sample tends to provide too low an estimate of the population standard deviation. Using $n-1$ is said to provide an *unbiased* estimate.

> **Exercise 7.4:** Calculate the standard deviation of the sample in Exercise 7.2

Let us consider our synthesis experiment again. If we repeat the synthesis several times we expect the measured yield to vary from time to time, even if the conditions are kept as fixed as possible. If we are only interested in seeing how large this variation is the standard deviation is a useful measure. This is because the standard deviation has the same unit as the yield and can be directly compared to it. In some situations it is interesting to obtain more detailed information about the variation. We might want to analyze different parts of the variation separately. For example, a part of the measured variation comes from the synthesis itself, since there will always be small variations in the concentrations of reactants, the homogeneity of the mixture, its temperature and so forth. All of these factors may affect the yield. We also know that another part of the variation comes from the measurement procedure, as the balance for weighing the product has a certain precision, the person weighing it has a certain repeatability and so on. Suppose that we are interested in comparing the part of the variation coming from the synthesis to the one coming from the measurement. In that case the standard deviation is not a suitable measure, because standard deviations cannot be added. Variances, on the other hand, can.

This property of variances can be understood using the Pythagorean theorem. Looking at Figure 7.5 we immediately see that the distance c is not the sum of the distances a and b, but Pythagoras tells us that $c^2 = a^2 + b^2$. The distance a is given by coordinates on the x-axis and b by coordinates on the y-axis. Since a and b are characterized by two

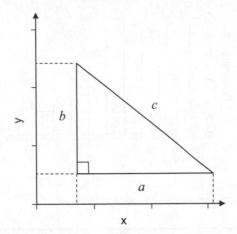

Figure 7.5
The sides of a right-angled triangle cannot be added to obtain the hypotenuse, but the Pythagorean theorem states that the sum of their squares equals the square of the hypotenuse. Similarly, standard deviations cannot be added, while variances can.

different variables (x and y) they can vary *independently* of each other. Now, returning to the synthesis experiment, the variability of the measurement system can be expected to be independent of the variability of the synthesis. This is because they are governed by different processes, so one can change without affecting the other. The total variance of the measured yield can thereby be written on the "Pythagorean" form:

$$\sigma_{total}^2 = \sigma_{synthesis}^2 + \sigma_{measurement}^2 \tag{7.5}$$

If we know the measurement precision and the total variation we can use this equation to determine the variability of the synthesis. Trying to add standard deviations would be like trying to add the sides of a right-angled triangle to obtain the hypotenuse. If the variance were to be divided into several independent components, Equation 7.5 could be generalized to as many dimensions as desired:

$$\sigma_{total}^2 = \sigma_1^2 + \sigma_3^2 + \cdots + \sigma_n^2 \tag{7.6}$$

Now, let us use what we have learned so far to try to analyze a practical problem. Do your best to answer the following exercises. We will return to them in the next chapter.

Exercise 7.5: Sven is the proud owner of a car with a diesel engine. He claims that the fuel consumption is less than 6.0 liters per 100 km. During the Scandinavian summer he filled the tank five times and calculated the following fuel consumptions in liters per 100 km:

5.9	5.8	5.9	6.2	6.1

Based on these measurements, would you say that he is right to claim that the real fuel consumption is less than 6.0 liters per 100 km?

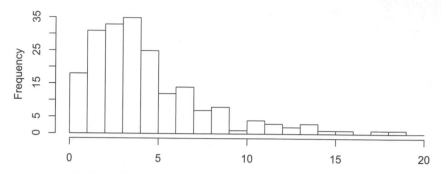

Figure 7.6
Histogram of a right-skewed sample.

Exercise 7.6: Sven also filled the tank five times during the Scandinavian winter and calculated the following fuel consumptions in liters per 100 km:

5.9	6.3	6.0	6.1	6.0

Is there a difference between the summer and winter fuel consumption?

Besides the measures introduced so far, we may also obtain an impression of the central tendency and variation of a sample using graphical tools. One such tool is the box plot, which was shown in Figure 7.2. Another common tool is the histogram. It provides a more detailed overview of the distribution of values in the sample than the box plot. An example is shown in Figure 7.6. Histograms are created by sorting the values from the smallest to the largest and dividing them into classes, or bins, covering intervals of equal width. The number of values in each bin is called the frequency. Each bin is represented by an interval on an axis and the frequencies by rectangles over the intervals. The area of each rectangle is proportional to the frequency of its bin. In Figure 7.6 we see that all the values in the sample are contained in an interval from zero to 20, with most of them residing near three. It is most likely to find a value in the bin covering the interval from three to four. Since this bin has the highest frequency it is called the *modal bin*.

Besides assessing the central tendency and variation of a sample, the histogram can be used to describe the distribution of the data. Most of the data in Figure 7.6 extend in a tail on the right-hand side of the modal bin. This distribution is therefore said to be right-skewed. A left-skewed distribution, correspondingly, has a tail extending to the left. If two similar tails extend in both directions from the modal bin the distribution is said to be symmetric. You have probably heard of the so-called *normal distribution*, which is one case of a symmetric distribution that is of general interest. It is discussed at some length in the next couple of sections. It is useful to remember that it requires rather a large amount of data to judge from a histogram if the data are normally distributed. Usually, 30–50 values are considered to be a reasonable minimum.

7.4 Probability Distribution

The assumption of random sampling means that we consider all observations in a sample to have been made with equal probability. Under such conditions, the probability of finding a value in a certain bin is simply its frequency divided by the sum of all frequencies in the histogram. The probability of finding any value whatever in the distribution is exactly one, as this is the sum of the probabilities of all bins.

Since the histogram describes the probability of finding values in a limited number of discrete bins it is called a discrete distribution. If we are interested in calculating probabilities of continuous variables we need a continuous distribution. Such distributions are called *probability density functions*. They are explained here using a bit of mathematics, since integrals and derivatives are useful for illustrating the ideas. For readers who have forgotten some of their calculus, we do not have to be able to calculate the integrals manually. As we shall see soon, common software has built-in functions that do this.

We can think of probability density functions as histograms with a very large number of infinitesimally thin bins. (An infinitesimal interval is an interval so small that, although it is not of exactly zero size, it cannot be distinguished from zero.) In contrast to histograms, probability density functions appear as smooth curves in diagrams. Just as before, the sum of the probabilities of all these infinitesimal intervals is exactly one and, with continuous variables, such a sum is represented by an integral. We will let the symbol dX represent an infinitesimal increment in the variable X. The probability that X attains any value at all can then be described by the integral:

$$\int_{-\infty}^{\infty} f(X)dX = 1 \tag{7.7}$$

where $f(X)$ is the *probability density function* of X. The integral corresponds to the area under the curve described by $f(X)$ between the values under and above the integral sign. In this case it is the area under the whole curve, from minus infinity to plus infinity. The probability P of finding the variable between the values a and b is given by the integral:

$$P(a \leq X \leq b) = \int_{a}^{b} f(X)dX \tag{7.8}$$

The probability of X attaining a value smaller than x is:

$$P(X \leq x) = F(x) = \int_{-\infty}^{x} f(X)dX \tag{7.9}$$

$F(x)$ is called the *cumulative distribution function* and corresponds to the area under $f(X)$ between minus infinity and x. Examples of a probability density function and its corresponding cumulative distribution function are shown in Figure 7.7. The cumulative distribution function increases steadily from zero to one and we may note that the probability density function peaks where the slope of the cumulative distribution function is greatest. This is because it is the derivative of the cumulative distribution.

Since these two functions essentially contain the same information we might ask why it is necessary to use both of them. A glance at the probability density function quickly gives an idea about the most probable value and the variation around it, but we cannot tell the probability of any specific value just by looking at the curve. Since it is a *density* function describing a continuous variable, the probability for any given value is actually infinitesimally small. To read out probabilities we need the cumulative distribution function. When sampling from a population described by the probability density function in Figure 7.7, the probability P of obtaining a value smaller than x can be read directly on the

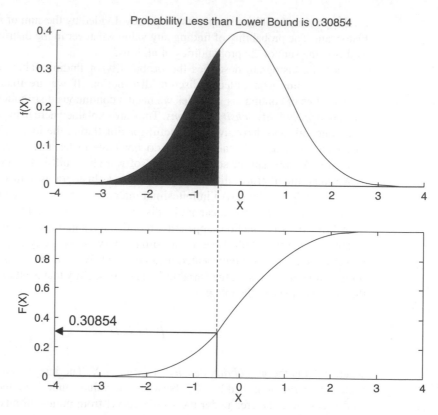

Figure 7.7
Top diagram: probability density function. The shaded area corresponds to the probability of finding a value $X < -0.5$. Bottom diagram: cumulative distribution function. The probability of $X < -0.5$ can be read out on the $F(X)$-axis.

$F(X)$-axis, as described in the figure. This number corresponds to the area of the shaded part under the probability density curve. The probability of obtaining a value greater than x is $1 - P$. This is a direct consequence of the probability of obtaining any value at all being exactly one.

The specific probability density function depicted in Figure 7.7 is called the *normal distribution* and is of central importance in statistics. It is also called the bell curve or Gaussian function, after the German mathematician Carl Friedrich Gauss who determined its formula. An important property of the normal distribution is that it is uniquely defined by its mean and standard deviation. It is useful because the random nature of many processes automatically generates this type of probability distribution. For example, repeated observations that differ due to experimental error often vary about a central value in a roughly symmetric fashion, with small deviations occurring much more frequently than large ones. When the number of observations becomes large, their frequency distribution tends to approach the normal distribution. The reason for this will become clearer when we discuss the central limit effect.

Readers who have forgotten how to determine integrals do not have to worry; these probabilities do not have to be calculated by hand. They are readily determined using common software. A version of Microsoft Excel® that has statistical functions is fully sufficient for many statistical analyses. If you use MathWorks MATLAB® and have access to the Statistics Toolbox, you are amply provided with statistical tools. If you do not have access to such software you may use statistical tables of common distribution functions, such as the ones in the appendix of this book.

Example 7.2: Let us assume that we are sampling from a normal distribution with mean 5 and standard deviation 3. What is the probability of drawing a value smaller than 4? In Excel this probability can be calculated using the worksheet function NORMDIST, which returns the normal distribution function. (Note that non-English versions of Excel may use other function names.) It requires four input arguments: the value for which the distribution is wanted (4), the mean (5), the standard deviation (3), and one specifying if we want the cumulative function or the density function. Type the following expression into a cell in an Excel worksheet: "=NORMDIST(4;5;3;TRUE)". When you press return you obtain the probability that we asked for, which is 0.369. The cumulative distribution is returned if the last argument is "TRUE". Similarly, the probability of obtaining a value greater than four is given by "=1-NORMDIST(4;5;3;TRUE)". ∎

It is often helpful to use a special case of the normal distribution, called the *standard* normal distribution. Any normally distributed variable can be described by the standard normal distribution if a simple variable transformation is made. Here, transformation means that the new variable contains the same information as the original one, but in different units. If X is a normally distributed variable with mean μ and standard deviation σ, then the transformed variable:

$$Z = \frac{X - \mu}{\sigma} \qquad (7.10)$$

follows the standard normal distribution. If we let X be equal to its mean, μ, we see that the mean of the new variable Z is zero. If we instead let X lie exactly one standard deviation from the mean ($X = \mu + \sigma$), we see that Z attains the value one. This means that the new variable Z follows a normal distribution with mean zero and standard deviation one. Since any normally distributed variable can be transformed into Z, they can all be analyzed using the standard normal distribution. For this reason, most tables over normal probabilities contain the standard normal distribution.

In the theoretical normal distribution, 68.3% of the data fall within one standard deviation from the mean. 95.4% fall within two and 99.7% fall within three standard deviations from the mean (Figure 7.8). With samples of limited size it is difficult to estimate the mean and standard deviation of the population. When analyzing real data it is therefore reasonable to expect about 60–75% of the values to fall within one standard deviation from the mean, 90–98% within two, and 99–100% within three. This means that, when we know the mean and standard deviation of a normally distributed sample, we can say something about the probability that a certain value comes from the same population.

It is common to plot the standard deviation as error bars in diagrams of measurement data. This is motivated by the measurement error often being normally distributed. Since essentially all the data in a small, normally distributed sample are expected to fall within three standard deviations of the mean, values outside this range are often called *outliers*. They might be considered suspect values, possibly arising through disturbances or mistakes during measurement. Two things are important to remember. Firstly, you should never remove outliers without good reason, since that would be tampering with the data to obtain a result that pleases you. Outliers indicate a need to investigate if the measurements have

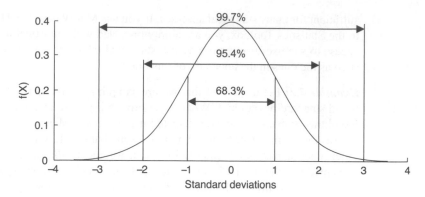

Figure 7.8
The portions of the data falling within one, two and three standard deviations in a theoretical normal distribution.

gone wrong, but remember that they can occur naturally, by chance. Secondly, if the error is not normally distributed, you cannot unreflectively use the standard deviation as a criterion for identifying outliers. Self-evident as this may seem, even established researchers sometimes assume that you can.

Example 7.3: Let us assume that the average caries rate in the nation as a whole is 14% and that the standard deviation between towns is 3.1%. The citizens of Caramel have a caries rate of 19%, which is well above the national average. If we assume that the caries rate is normally distributed among towns we may estimate how common it is to find a caries rate as high as the one in Caramel. About 70% of the towns are expected to be within one standard deviation from the mean, or between 10.9 and 17.1%. About 95% of the towns are expected to be within two standard deviations from the mean, or between 7.8 and 20.2%. Since the caries rate in Caramel is within this interval, more than one out of twenty towns can actually be expected to have a caries rate that is farther from the national average than that in Caramel. ∎

Exercise 7.7: Let the caries rate of a town be described by the normally distributed variable X. The mean caries rate μ is 14% and the standard deviation σ is 3.1%. In Caramel, $X = 19\%$. Calculate the Z-score for Caramel using Equation 7.10. Calculate the probability for obtaining a caries rate equal to or greater than that in Caramel. This can be done using the table of standard normal probabilities in the Appendix. In Excel, it is done using the function NORMSDIST(Z), which returns the standard normal cumulative distribution function of Z. Use Figure 7.7 to interpret this in terms of probabilities.

7.5   The Central Limit Effect

How can it be that the normal distribution holds such a central position in statistics? There are many other ways that data may be distributed. If we roll a six-sided die, for example, the probability of each score is 1/6. In other words, the frequency distribution of the scores is completely flat, far from the bell-shaped Gaussian curve. The roll of a die is a relatively simple process, however. Measurements are more complex. The error in measurement data is usually combined from a large number of component errors. When discussing the synthesis experiment earlier, we mentioned a few sources of error that could contribute to the overall error: variations in reactant concentrations, mixture homogeneity, temperature, as well as measurement precision and operator repeatability. There are numerous other potential error sources, such as the purity of the reactants. We do not know which distributions these individual errors follow but due to the *central limit effect* we know something about how the combination of a large number of such errors is distributed. This effect makes the mean of a number of random variables follow a normal distribution when the number of variables becomes large, almost regardless of how each variable is distributed.

The most simple and effective way to demonstrate the central limit effect is perhaps by tumbling dice. As we have mentioned, we expect the frequency distribution of scores from tumbling a true die to be flat. The top histogram in Figure 7.9 shows the result of 600 throws of a single die. The frequency of each score is approximately 100, or one sixth of the number of throws, just as expected. The mean score is 3.5. What happens if we throw two dice simultaneously and calculate the *average* score from each throw? The middle histogram shows that the distribution now takes on a pointed appearance; it has lost its flatness. Intuitively, this can be understood by noting that there is only one combination of scores from two dice that produces the lowest possible mean score, which is one. The same is true about the highest possible score, six, whereas there are a large number of combinations that produce a mean score close to 3.5. Since each combination occurs with

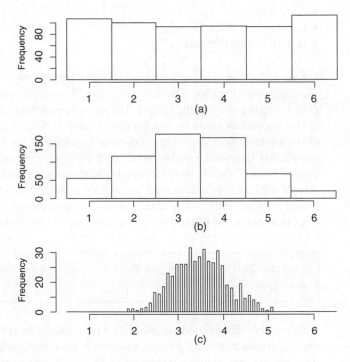

Figure 7.9
Histograms showing the scores from throwing a single die (top), two dice (middle) and ten dice (bottom). (When several dice are thrown the mean score is calculated.)

the same probability the frequency becomes highest around the mean. Looking at the bottom histogram, which results from throwing ten dice simultaneously and calculating the mean score, we see that the distribution approaches the Gaussian curve. With so many dice the probability of each unique combination becomes so low that 600 throws were not enough, in this case, to produce a single occurrence of neither the lowest nor the highest possible score.

This demonstrates how the mean of a large number of random variables tends to follow a normal distribution, even though their *individual* distributions may deviate significantly from normality. This is a remarkable and highly useful effect. The reason that the normal distribution is so important is that many natural processes are affected by a wealth of factors and subprocesses. Also, our measurement systems are affected by a large number of random errors. In real-world experimentation, therefore, it is often reasonable to assume that the experimental error follows a normal distribution. An important condition for this to be true is that the contribution from each component of the variation is of the same order of magnitude. If one component dominates over the others we can no longer adduce the central limit effect and assume that the error is normally distributed.

The central limit effect is stated mathematically by the *central limit theorem*, which will be used for calculating confidence intervals later in this chapter. Let us consider a variable X that follows an unknown distribution with mean μ and standard deviation σ. Imagine that we take samples of Xs, each containing n observations. The central limit theorem states that, when n becomes large, the *means* of these samples will tend to follow a distribution that:

- has mean μ;
- has standard deviation $\sigma/\sqrt{n}$;
- is normally distributed.

Let us reflect on what this means. The last point is the central limit effect, which we illustrated with the dice. The first point states that we can estimate the population mean by calculating the mean of the sample. The second point states that we obtain a *better* estimate of the population mean by using a *larger* sample. This is because the standard deviation of the estimate decreases when n increases, exactly as we saw in the dice data: when the sample size increased from one to two, and later to ten, the estimated mean remained 3.5 but the *precision* of the estimate increased. The decrease in standard deviation with increasing sample size reflects a decreased uncertainty in the estimate. I suggest that you read this paragraph repeatedly until you are sure that you understand what it says, because the central limit theorem is one of the most important things to understand in statistics.

Exercise 7.8: Why is it reasonable to assume that the caries rate is normally distributed among the towns of the nation?

Remember that the normal distribution is an ideal concept – in the real world we seldom encounter data that exactly follow this distribution. Fortunately, most statistical techniques only require approximate normality to be useful. For example, the techniques for estimating and comparing means that will be introduced in the next chapter are robust to non-normality.

7.6 Normal Probability Plots

It can be difficult to see if data are normally distributed in a histogram, especially if the sample is small. Normal probability plots, or normal plots for short, are complementary graphical tools that can be used for this purpose. Apart from checking the normality of the data they can also point to individual values that do not originate from a normal distribution. This is why they will be important when we introduce experimental design in Chapter 9. In experimental data, some part of the variation is due to measurement error and some part to our deliberate experimentation. We hope that at least some effects of the experiment are greater than the measurement noise. Such effects do not follow the normal distribution of the error and normal plots can thereby be used to detect them.

In normal plots the data are essentially plotted against a theoretical normal distribution. If the data are normally distributed they will plot on an approximately straight line. Figure 7.10 shows such an example. The *x*-axis represents the data in the sample, roughly scattered between zero and 35 with most values in the middle of this interval. The *y*-axis represents the cumulative probabilities of a theoretical normal distribution. To aid the eye, the figure also contains a straight line drawn through the upper and lower quartiles. Figure 7.11 shows a normal plot of a right-skewed sample. Recall that the distribution of such a sample has a tail extending to the right. Since the frequency of low values in this sample is higher than would be expected from a normal distribution, the slope of the data is steeper on the left and shallower on the right. Conversely, a left-skewed sample would be steeper on the right.

Since it is important to understand the principles of the normal plot we are going to look at how the cumulative probabilities on the *y*-axis are determined. There are thirty observations in the sample plotted in Figure 7.10. Imagine that you have thirty intervals on an axis. Sort

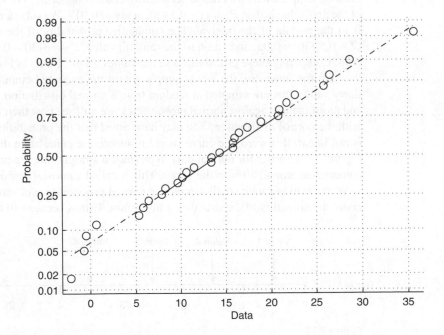

Figure 7.10
Normal plot of a normally distributed sample.

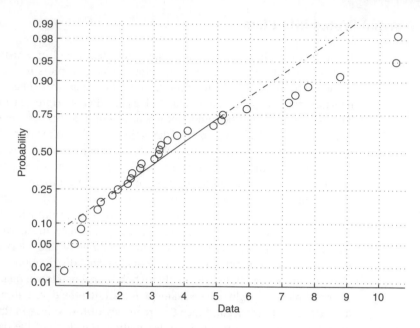

Figure 7.11
Normal plot of a right-skewed sample.

the observations from the smallest to the largest and put them in the middle of these intervals, as shown in Figure 7.12. Each interval now represents one thirtieth of the data. If it is a random sample each observation occurs with equal probability. The cumulative probability of the first value is then $P_1 = (1 - 0.5)/30$, or about 0.017. We subtract 0.5 because the value is in the middle of the interval. The cumulative probability of the second value is $P_2 = (2 - 0.5)/30 = 0.05$, and so on to the thirtieth value, $P_{30} = (30 - 0.5)/30$, which is about 0.98. Comparing these probabilities to the values on the y-axis in Figure 7.10, we see that they are the same. So, the y-axis simply shows the expected cumulative probabilities of thirty values that are sampled at random from a normal distribution. By ordering our data and plotting them against normal probabilities, we will see how their distribution compares with the normal distribution. You may have noted that the probability scale in Figure 7.10 is not linear. If it were, the curve would resemble the cumulative distribution function in Figure 7.7. In order for normal data to produce a straight line the probabilities have been compressed around 50% on the y-axis. This is called a *normal probability scale*.

When working with small samples, a normal probability plot often gives a better idea about the normality of the data than a histogram. This is because all the values are plotted,

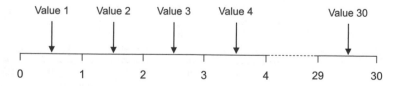

Figure 7.12
Sorting values in a sample, from the smallest to the largest, into intervals on an axis to determine their expected cumulative probabilities.

whereas a histogram collects the values into bins. Still, a reliable assessment of the normality of a sample requires a large sample.

Example 7.4: Since we are going to use normal plots in later chapters it is important to learn how to make them. Actually, making one will also improve our understanding of how they work. It is recommended that you carry out all the steps in this example yourself. It is easiest done using computer software with statistical functions. I am using Excel here, as most readers are likely to have access to this software, but the corresponding functions are of course available in MATLAB and other packages. The exercise can also be carried out using the table of standard normal probabilities in the Appendix.

Let us look at the following sample, drawn from a normal distribution:

8.6	12.5	0.2	9.6	7.9	3.0	5.7	8.0	17.7	15.3

The first step is to sort the data from the smallest to the largest value. Do this by writing them into a column in an Excel worksheet, selecting the values and then clicking the button "Sort A-Z". The ordered values are used as *x*-values in our plot.

Next, we want to obtain something that represents our cumulative probabilities, preferably without needing to generate the exotic normal probability scale. You may recall that the standard normal cumulative distribution function translates a normally distributed variable Z on the *x*-axis into cumulative probabilities on the *y*-axis, as the arrow in Figure 7.7 indicates. The *inverse* of this function goes from the *y*-axis to the *x*-axis (imagine reversing the arrow in Figure 7.7 to go from cumulative probabilities back to the Z-variable). Z is, of course, normally distributed with values more densely scattered around the mean and sparser toward the edges. So, plotting a normally distributed sample against Z in a clever way will produce an approximately straight line.

The *inverse* of the standard normal cumulative distribution is easily obtained using the Excel function NORMSINV(*P*). *P* is a cumulative probability and the returned Z-value is called a *normal score*. We can use this function to produce our normal plot as follows. Each value in our sample should first be assigned an index by writing numbers from one to ten in the column next to the ordered data. This corresponds to sorting them into intervals as in Figure 7.12. From these indexes we calculate cumulative probabilities as explained above, by writing a worksheet function into the cell next to the index. If the first index is in cell B1, the expression "= (B1 − 0.5)/10" should be written into cell C1. When we press RETURN the calculated probability, 0.05, occurs in place of the formula. The formula can then be applied to all the indexes by selecting the cell that contains the formula and simply double-clicking its fill handle. Finally, we want to calculate the normal scores. Write the expression "=NORMSINV(C1)" into cell D1 and apply it to all the probabilities in column C by double-clicking the fill handle. We should now have a table with the following values:

0.2	1	0.05	−1.644853627
3	2	0.15	−1.036433389
5.7	3	0.25	−0.67448975
7.9	4	0.35	−0.385320466
8	5	0.45	−0.125661347
8.6	6	0.55	0.125661347
9.6	7	0.65	0.385320466
12.5	8	0.75	0.67448975
15.3	9	0.85	1.036433389
17.7	10	0.95	1.644853627

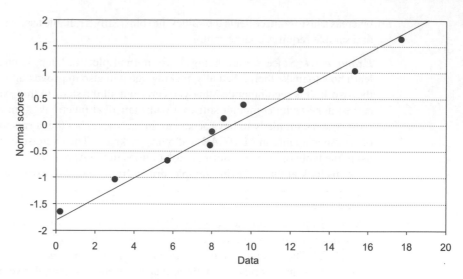

Figure 7.13
Normal plot of the sample in Example 7.4.

The first column contains the ordered data, the second their indexes, the third their cumulative probabilities, and the last contains their normal scores. The last step is simply to plot the ordered data against the normal scores (Figure 7.13). The line is drawn through the upper and lower quartiles, which are easily obtained using the worksheet function QUARTILE.

Note that the normal scores are measured in standard deviations. Reading the values on the *x*-axis that correspond to the locations where the normal scores equal plus or minus one, we see that they are about five units from the mean. Quite right, the calculated standard deviation of the sample is 5.3, as obtained using the worksheet function STDEV. ∎

7.7   Confidence Intervals

Let us return to our research team that wants to investigate how a certain diet affects the blood pressure of healthy people. Since they cannot test the diet on the whole population of healthy people they have chosen a representative sample consisting of 15 persons, which is a common sample size for these types of investigations. Before the tests they measure the blood pressures of their test persons. After 15 days of diet they measure the blood pressures again. Calculating the difference between the systolic blood pressure after and before the diet for each test person, they obtain the following results (in mm Hg):

−3	−5	−12	−4	−6	−14	−14	−4
+2	−6	−9	0	−5	−3	−4	

There seems to be a large variation in the results. Two persons' blood pressures drop by as much as 14 mm Hg, but in one person there is actually an increase. What can we say about the effect of the diet based on these figures? As a first step it is always useful to display the data graphically in a diagram. Figure 7.14 shows that the drop after the diet is about 4 mm Hg for most persons. A few persons show larger drops than this, while others show no

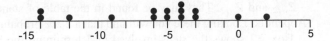

Figure 7.14
Differences in blood pressure before and after diet.

effect or even an increase. The mean difference in the sample is −5.8 mm Hg. But this is a limited sample. The interesting question is what the mean difference in blood pressure would have been in the whole population. Is it possible to be, say, 95% certain that the diet is effective?

Stating our problem in general terms, we have a sample from an unknown population and want to estimate a population parameter at a certain level of confidence. The interesting parameter in our case is the mean difference in blood pressure resulting from the diet and we want to have 95% confidence in our estimate of this parameter. Recall from the central limit theorem that, for large sample sizes n, the mean $\bar{x}$ of a sample approximately follows a normal distribution with mean μ and standard deviation $\sigma/\sqrt{n}$. Using Equation 7.10 we may thereby make the variable transformation:

$$Z = \frac{\bar{x} - \mu}{\sigma/\sqrt{n}} \qquad (7.11)$$

and obtain a variable that follows the standard normal distribution. Since this distribution is known it is now easy to find an interval such that there is 95% probability of finding Z within it.

Alternatively, we could say that there should be 5% probability of finding Z *outside* the interval; this probability is commonly written α. Looking at Figure 7.15 we see that α corresponds to the shaded areas under the tails of the curve. Each of the two areas corresponds to 2.5% probability, or $\alpha/2$. The critical values of Z that limit the interval are, therefore, called

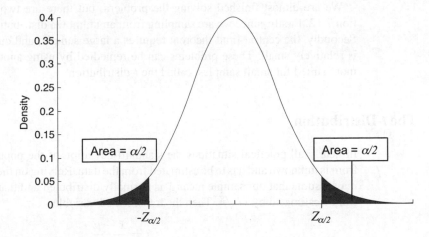

Figure 7.15
There is 95% probability of finding Z in the white region. There is thus 5% probability of finding Z in the shaded regions. This probability is called α. Since each shaded region corresponds to $\alpha/2$, the critical values limiting the 95% interval are called $Z_{\alpha/2}$.

$-Z_{\alpha/2}$ and $Z_{\alpha/2}$. They may be found in the table of standard normal probabilities in the Appendix. Looking there you will find that a 2.5% probability corresponds to $Z_{\alpha/2} = 1.96$.

Box 7.2 shows the steps involved in determining the 95% confidence interval for the population mean μ. The first step states that, by definition, there is 95% probability of finding Z between $-Z_{\alpha/2}$ and $Z_{\alpha/2}$. In the next step we replace Z using Equation 7.11, to get an expression that contains the population mean μ. Finally, we rearrange the terms to obtain an interval for μ. We arrive at the following expression for the 95% confidence interval for μ:

$$\mu = \bar{x} \pm Z_{\alpha/2}\frac{\sigma}{\sqrt{n}} \tag{7.12}$$

Box 7.2 Steps for obtaining a confidence interval for the population mean μ from the standard normal distribution

There is 95% probability that:

$$-Z_{\alpha/2} < Z < Z_{\alpha/2}$$

Replacing Z using Equation 7.11, we get:

$$-Z_{\alpha/2} < \frac{\bar{x} - \mu}{\sigma/\sqrt{n}} < Z_{\alpha/2}$$

This is equivalent to:

$$\bar{x} - Z_{\alpha/2}\frac{\sigma}{\sqrt{n}} < \mu < \bar{x} + Z_{\alpha/2}\frac{\sigma}{\sqrt{n}}$$

The meaning of this is that, if we take repeated samples and calculate the confidence intervals from them, we expect these intervals to include μ 95% of the time.

We are almost finished solving the problem, but there are two problems with Equation 7.12. Firstly, since we are sampling from an unknown distribution, we do not know σ. Secondly, the central limit theorem requires a large sample, and our sample of 15 people is relatively small. These problems can be remedied by using another distribution that is more suited for small samples, called the t-distribution.

7.8 The t-Distribution

In nearly all practical situations the standard deviation of the population that we sample from is unknown and has to be estimated from the data. Relying on the central limit effect we may assume that our sample mean $\bar{x}$ is normally distributed and that its standard deviation can be estimated by $s/\sqrt{n}$. Then the transformed variable:

$$t = \frac{\bar{x} - \mu}{s/\sqrt{n}} \tag{7.13}$$

follows a so-called t-distribution. As previously mentioned, a larger sample gives a more reliable estimate of the standard deviation, so you are probably not surprised that the

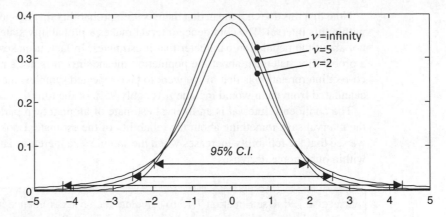

Figure 7.16
Examples of *t*-distributions with two, five and infinite degrees of freedom.

t-distribution depends on the number of degrees of freedom, ν, of the sample. This is illustrated in Figure 7.16, showing *t*-distributions for $\nu = 2$, 5, and infinity. Note that they become wider and flatter when ν decreases, but the mean is always zero. The *t*-distribution's standard deviation is:

$$\sigma = \sqrt{\frac{\nu}{\nu - 2}} \qquad (7.14)$$

meaning that, as ν becomes large, σ approaches one and the distribution approaches the standard normal distribution. The standard normal distribution is, therefore, a special case of the *t*-distribution.

Now, let us return to the dietary experiment. We did not know σ and had a sample too small for using the standard normal distribution. Now that we know of the *t*-distribution, which is suited for small samples, we may use Equation 7.13 and continue our analysis. The confidence interval for μ is simply:

$$\mu = \bar{x} \pm t_{\alpha/2, \nu} \frac{s}{\sqrt{n}} \qquad (7.15)$$

We arrive at this equation exactly as we did with Equation 7.12. (It is a good exercise to do this by working through Box 7.2 using *t* instead of *Z*.) Instead of $Z_{\alpha/2}$ we use $t_{\alpha/2, \nu}$, which must be determined for the desired significance level α, and the appropriate number of degrees of freedom ν, in order to calculate the confidence interval. In our sample, $\nu = 15 - 1 = 14$ and, for a 95% interval, $\alpha/2 = 0.025$. If you look in the table of probability points for the *t*-distribution in the Appendix you find that this corresponds to $t_{\alpha/2, \nu} = 2.145$.

In our sample, the mean difference in blood pressure is -5.8 mm Hg and the sample standard deviation is 4.7 mm Hg. Putting these values into Equation 7.15 we find that the 95% confidence interval for the mean blood pressure drop extends from -8.4 to -3.2 mm Hg. We may conclude that, at the 95% confidence level, the diet is effective.[1]

---

[1]Strictly speaking, this experiment would need a control group to ensure that the effect is due to the diet and not to a background factor. For further discussion, see the section "Reflections on the exhibition" in Chapter 6 and "Designs with one categorical factor" in Chapter 9.

Note that this does *not* mean that there is 95% probability that the population mean μ lies within this interval. The confidence interval makes a probability statement about the data, not about the population parameter that it estimates. In fact, it makes little sense to make a probability statement about the population mean, since it is not a random variable. The correct interpretation is that, if we were to take repeated samples, the confidence intervals calculated from them would include μ roughly 95% of the time.

The confidence interval is an interval estimate of an unknown parameter. The width of the interval says something about the reliability of the estimate. Looking at Equation 7.15 we see that the reliability increases when the sample size increases and when the variation within our sample decreases.

> **Exercise 7.9:** Use Equation 7.15 to calculate the 95% confidence interval for the real fuel consumption of Sven's diesel car during summer. The fuel consumption readings are given in Exercise 7.5.

In this chapter, we have introduced some important concepts that form the basis of more advanced statistical techniques. We have already seen how sample statistics can be used to estimate properties of the parent distribution that the sample came from, such as confidence intervals for the population mean. This type of analysis is called inferential statistics and is introduced further in the next chapter, where it will become clear why knowledge in statistics is useful in the design and analysis of experiments.

7.9   Summary

- Statistics is a way of coping with the variation that exists everywhere and for stating, with confidence, how uncertain we are about our conclusions.
- Methods for summarizing essential features of data are called descriptive statistics. They include measures of central tendency and variation, as well as graphical tools for depicting data.
- Populations consist of every possible observation and are generally impossible to obtain. In practice, we always use samples of limited size to represent the population. The population and sample *standard deviations* are denoted σ and s, respectively. The population and sample *means* are denoted μ and $\bar{x}$, respectively.
- Random sampling means that every possible observation of a certain condition occurs with equal probability. Since random sampling is a central assumption in statistical analysis it is of great importance that we collect data in a way that produces random samples. This is often more difficult than it seems.
- A probability density function $f(x)$ describes the probability of finding a random variable in an infinitesimally thin interval between the values x and $x + dx$. This probability is equal to $f(x)dx$.
- A cumulative distribution function $F(x)$ describes the probability that a random variable is smaller than the value x. $F(x)$ increases with x and can attain values between zero and one.
- The standard normal distribution has mean zero and standard deviation one. Since any normally distributed variable can be transformed into one that follows the standard normal

distribution, it is the only distribution we need to analyze normally distributed data. Most tables of normal probabilities are based on the standard normal distribution.

- When sampling from an unknown population that has mean μ and standard deviation σ we expect, for sufficiently large sample sizes, n, that the sample means will tend to follow a normal distribution with mean μ and standard deviation $\sigma/\sqrt{n}$. This is the central limit effect and it explains why the normal distribution is of such general value in statistics.
- In normal probability plots, the values of a sample are plotted against a theoretical normal distribution. A normally distributed sample will thereby plot approximately on a straight line.
- Due to sampling error, there is some uncertainty involved in estimating a population parameter from a sample of limited size. It is often useful instead to estimate an interval that has a specified probability of containing the parameter. Such intervals are called confidence intervals.
- The t-distribution is a family of normal-shaped distributions that are used for estimating the mean of a population from a small sample. The standard deviation of the t-distribution decreases when the sample size increases.

Further Reading

The contents of this chapter are described in most elementary textbooks on statistics. Many such books take a rather mathematical approach to the subject. It is often beneficial for the understanding to present statistical ideas and tools by practical examples. *Statistics for Experimenters* by Box, Hunter and Hunter [1] does so in a clear way and is probably one of the better statistics books available for people doing experimental work. Those readers who want to write programs for statistical analysis might consider *Statistics – an introduction using R* by Crawley [2]. "R" is a freely available language and environment for statistical computing and graphics. It is platform independent and can be downloaded at http://cran.r-project.org/.

Answers for Exercises

7.1 Make a numbered list over all addresses in the nation. Draw the numbers using a random number generator, for example using the Excel function = RAND().

7.2 Mean = 7.22, median = 7, mode = 7.

7.3 Mean = 3, median = 3 for all three dot plots. The range is 0 in the top dot plot and 4 in the other two.

7.4 Standard deviation = 3.96.

7.5 See the answer for Exercise 8.9 and the discussion of these data in connection to the one-sample t-test in Chapter 8.

7.6 See the discussion of these data in connection to the two-sample t-test in Chapter 8.

7.7 $Z = 1.61$. The cumulative probability $F(Z) = 0.9463$ (read from the table of standard normal probabilities in the Appendix). The probability of finding a rate greater or equal to that in Caramel is $1 - F(Z) = 0.0537$. A more exact value can be calculated in Excel using the expression = 1-NORMSDIST(Z).

7.8 According to the central limit theorem, the means of samples tend to be normally distributed for large sample sizes. The caries rate in a town is the mean of a sample consisting of all its inhabitants. Since the central limit theorem applies, it is reasonable to expect a normal distribution.

7.9 Reading from the table over probability points for the *t*-distribution in the Appendix, $t_{0.025, 4} = 2.776$. The confidence interval lies between 5.8 and 6.2 liters per 100 km.

References

1. Box, G.E.P., Hunter, J.S., and Hunter, W.G. (2005) *Statistics for Experimenters: Design, Innovation, and Discovery*, 2nd edn, John Wiley & Sons, Inc., Hoboken (NJ).
2. Crawley, M.J. (2005) *Statistics: An Introduction Using R*, John Wiley & Sons Ltd, Chichester.

8 Statistics for Experiments

It is possible, and indeed it is all too frequent, for an experiment to be so conducted that no valid estimate of error is available. In such a case the experiment cannot be said, strictly, to be capable of proving anything. Perhaps it should not, in this case, be called an experiment *at all, but be added merely to the body of* experience *on which, for lack of anything better, we may have to base our opinions.*

—Ronald A. Fisher

In this chapter we will look at statistical techniques that are common in many types of experiments. Many readers have probably heard about them but they are treated here because it is important for experimenters to understand how they work. An obvious reason is that it is difficult to interpret statistical results if you do not understand these techniques. Another reason is that it often helps to plan experiments so that they *can* be analyzed using statistical techniques. The foundations of the techniques were given in Chapter 7 and different aspects of planning experiments will be treated in subsequent chapters.

Keep in mind that you cannot learn any technique by reading about it in a book. Knowledge cannot be transferred – it comes about through an inner, intellectual process. Turning information into practical knowledge requires an active effort on your part. Although you may later use statistical software to analyze your data, you should complete the exercises in this chapter to make sure that you have not misunderstood details of the techniques. Many exercises can be solved manually and, in other cases, common software is sufficient to complete the tasks. The best way to learn statistics is to apply it. For this reason you are encouraged to actively think about ways to apply the techniques in this chapter to your own research.

It is important to point out that this chapter, and indeed this book, is too short to provide a comprehensive treatment of statistical techniques. I have simply chosen a few of the most common techniques in experimental science and tried to present them with less mathematics than many textbooks on statistics do. Further reading is suggested at the end of the chapter for those who want a deeper theoretical treatment of the subject. A good, general recommendation is to always keep a real statistics book within reach on the bookshelf.

8.1 A Teatime Experiment

This example is borrowed from Fisher's classical book [1], as it presents some important ideas in an everyday context that requires no previous knowledge. Once you understand

Experiment!: Planning, Implementing and Interpreting, First Edition. Öivind Andersson.
© 2012 John Wiley & Sons, Ltd. Published 2012 by John Wiley & Sons, Ltd.

this simple example you will also understand why statistics is useful in the planning and interpretation of many experiments.

Imagine a lady who likes to take milk with her tea and claims that she can taste whether the milk was added to the cup before or after the tea. Our aim is to design an experiment to test this ability. We explain to her that we are going to prepare eight cups of tea. The milk will be added first to four of them, and after the tea to the other four. These two ways of preparing the tea will be called two *experimental treatments*. We will then present the cups to her in random order and ask her to find the four where the milk was added first.

Even if she does find the right cups, this could of course be a lucky coincidence. It is therefore useful to think about the all the possible outcomes and decide how to interpret them. Since her task is to choose four cups to put in one category, leaving the remaining four in the other, this particular experiment has 70 possible outcomes. This is because the first cup can be chosen in eight different ways, the second in seven ways and so on, yielding $8 \times 7 \times 6 \times 5 = 1680$ possible sets. However, many of these sets will contain the same cups, only arranged in different order. There are $4 \times 3 \times 2 \times 1 = 24$ ways of arranging a set of four cups, as the first can be chosen in four ways, the second in three and so on. Dividing 1680 by 24 we get 70 possible outcomes.

Out of these outcomes, only one is completely correct. A person that entirely lacks the ability to taste the difference between the treatments is thereby expected to make a correct classification once in about 70 trials.

In how many ways can she get three cups right? There are four ways to pick three cups from the correct set, and four ways to pick the remaining cup from the wrong set. This means that there are $4 \times 4 = 16$ ways to get three cups right. If our lady lacks the ability to discriminate between the treatments, she is still expected to get three cups right in 16/70 of the trials.

By similar reasoning we may find that there are 36 ways of getting two cups right, 16 ways of getting 1 cup right, and only one way of getting none of them right. This defines the frequency distribution that the experiment follows in the case that the outcome is completely governed by chance. It is shown graphically in Figure 8.1 and, no, it is *not* called the "tea" distribution!

Now, would the interpretation of the experiment be different if we had used only six cups? With three cups belonging to each treatment the experiment would have 20 possible

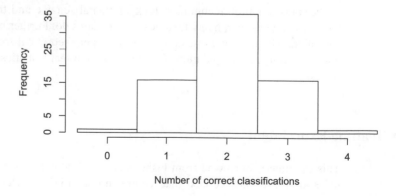

Figure 8.1
Frequency distribution of the possible outcomes of the teatime experiment. This is the expected distribution if the outcome were governed completely by chance.

outcomes. As one of them is correct, 5% (one in twenty) of the trials is expected to produce the correct classification if the experiment is governed completely by chance. As we mentioned in the last chapter, experimenters often choose a confidence level of 95% to judge if a result is statistically significant. An experiment utilizing six cups is thereby not capable of producing a significant result. With eight cups, however, the probability of obtaining the correct result purely by chance is 1/70. Being much less than 5% this result is significant at the 95% confidence level.

This analysis helps us realize that no experiment is capable of *proving* the existence of any phenomenon, because there is always a slight chance that the "correct" result occurs by chance. You may recall from Chapter 2 that one of the problems with the inductive method is that we cannot logically prove a statement, regardless of how many observations we have to support it. We cannot exclude the possibility that a future observation may disprove it. Instead of aiming to *prove* our conclusions we should use an experimental procedure that is capable of providing a *statistically significant* result. This means that the probability that the result has occurred purely by chance falls below a specific limit, called the significance level.

Let us say that our lady claims that, although she has the ability to taste the difference between the treatments, she does not expect to be 100% right. How should we interpret the outcome if she classifies three cups correctly? The frequency distribution tells us that the probability of getting three or more cups right by chance is 17/70 (there is one way of getting four cups right and 16 ways of getting three right). This is more than 5% and, consequently, such a result cannot be considered to be statistically significant. If we want to improve the *sensitivity* of the experiment, allowing our lady to make a mistake, we have to increase the *size* of the experiment. For example, with twelve cups (six belonging to each treatment) the probability of classifying at least five cups correctly can be calculated to be 37/924, which is less than 5%.

The test of significance is the central element when interpreting the experimental result. It divides the outcomes into two classes: those that support the hypothesis that she lacks the ability to taste the difference between the treatments and those that support her having this ability. The first hypothesis is called the null hypothesis. It assumes that the outcome of the experiment is governed by chance and is directly coupled to the distribution of possible outcomes in Figure 8.1. The null hypothesis cannot be proven by our data, but it may be rejected on the basis of our significance test. The central idea behind this type of statistical analysis is that the experiment is conducted *only to give the facts a chance to reject the null hypothesis*. In fact, this idea is the statistician's version of the hypothetico-deductive method introduced in Chapter 2: if the data do not support our hypothesis we reject it, otherwise we keep it and subject it to further tests.

8.2 The Importance of Randomization

If the lady should lack the ability to distinguish between the treatments, the experiment must be governed by chance for the significance test to be useful. As we mentioned in the last chapter, there are always numerous factors in experiments that are more or less outside our control, which may still affect the outcome. In this example, the quantity of milk might differ between the cups, the strength of the tea and the tasting temperature may change over time, and so forth. All of these factors could cause variations in taste that are greater than that caused by the experimental treatment. If, for instance, different persons were to prepare the cups belonging to the different treatments, systematic variations in taste could be introduced that were not due to the treatments. We could go to considerable lengths trying to eliminate such disturbances, inventing ever more accurate methods of pouring the milk and brewing the tea, but it would be a futile exercise to try to do away with them completely. It simply cannot be done.

It is important to keep in mind that experiments are always conducted using limited resources. When planning an experiment we must consider how to best utilize these resources. Which causes of disturbance should we care about and which *ought* to be deliberately ignored? How much effort should be put into diminishing the magnitude of those disturbances that we choose not to ignore? Most importantly, since it is impossible to eliminate the disturbances completely, it is important to *randomize* the experiment as in this example, where we present the cups to the lady in random order. Randomization minimizes the risk of systematic errors due to drift in the conditions. It maintains the integrity of the frequency distribution, which is the basis for evaluating the results. In short, randomization is what makes the significance test valid [1].

It should be said that it is sometimes difficult or even impossible to randomize an experiment. Luckily, there are techniques for dealing with such situations. Repeated replication of one or more measurement points over time is one way of detecting drift in the conditions. It is also possible to run the experiments in such an order (the run order) that makes it possible to separate experimental effects from time trends. If you find yourself unable to randomize your experiment you are recommended to consult a statistics book or, better still, a statistician. In most cases, however, randomization should be standard procedure.

8.3   One-Sided and Two-Sided Tests

In the teatime experiment we would discard the null hypothesis only if the lady had classified all four cups correctly. We call this a one-sided test, because only observations on one side of the reference distribution are capable of rejecting the null hypothesis. Classifying four cups correctly is the only "correct" outcome that would occur with a probability of less than 5% under the null hypothesis. Getting none of the cups right is equally improbable but that outcome would, of course, support the null hypothesis.

If we were to test the assertion that plants grown from a batch of seeds reach a certain height at a certain time after sowing, we would be making a *two-sided* test. The null hypothesis would be that they do reach the specified height, allowing for some random variation. It would be discarded if the plants were either too low or too high. Assuming that the height is expected to follow a normal distribution, this means that observations in two regions under the tails of the normal curve would weaken the null hypothesis. Testing at the 95% confidence level the combined area of these two tail regions would be 5%, meaning that each corresponds to 2.5% probability.

The reason that it is important to be aware if you are making a one- or two-sided test is that most tables of distribution functions assume a one-sided test. We already encountered this problem when calculating confidence intervals in Chapter 7, as the table of the t-distribution in the Appendix is based on this assumption. It states that the 5% point ($\alpha = 0.05$) for four degrees of freedom is $t = 2.132$. This means that there is 5% probability of finding a t-value greater than 2.132 (or smaller than -2.132, since the distribution is symmetric about the mean). For a two-sided test we would have to look for the $\alpha/2$ point, which is 2.776. This is because the combined probability of finding a value that is *either* smaller than -2.776 *or* greater than 2.776 is 5%.

Table 8.1 One- and two-sided probabilities from the *t*-distribution. Italicized *t*-values are correct for a two-sided test at the 95% confidence level, shaded values are correct for a one-sided test.

α	*t*-table	TINV
0.025	2.776	3.495
0.05	2.132	2.776
0.1	1.533	2.132

A word of caution is necessary at this point, as this book shows how to solve statistical problems in Microsoft Excel®. To add confusion to the situation, the Excel worksheet function TINV (which provides the probability points of the *t*-distribution) assumes that we are making a *two-sided* test. Table 8.1 is provided to help alleviate this confusion. It gives the probability points for the *t*-distribution with four degrees of freedom as returned from the table in the Appendix and from the TINV function. The α-values in the first column are the entry values of the table and the function. The italicized *t*-values are correct for a two-sided test at the 95% confidence level, whereas the shaded values are correct for a one-sided test.

8.4 The *t*-Test for One Sample

Let us revisit Exercise 6.5, where Sven claimed that the fuel consumption of his diesel car was less than 6.0 liters per 100 km. In the summer, he measured the following fuel consumptions in liters per 100 km:

5.9	5.8	5.9	6.2	6.1

In the exercise we were asked to determine if his claim was right. When I give this exercise to students after a lecture covering descriptive statistics, similar to how it was presented in Chapter 7, most course participants approach the exercise using descriptive statistics. They try to use the mean and standard deviation of the sample to answer the question. Typically, one half of a class concludes that Sven is right since the mean fuel consumption is below 6.0. The other half argues that, since some values are above 6.0, his claim must be incorrect. Although many students are hesitant to draw a definite conclusion, most are leaning towards one of these alternatives. They seldom claim that the question is wrongly posed.

Is it reasonable that researchers who are faced with the same problem and the same data arrive at completely different conclusions? Of course not, since the differences result from arbitrary and subjective considerations. After discussing this situation, some students revise their opinion and say that these data are *insufficient* to draw a conclusion.

Firstly, we might ask ourselves which number of measurements would be sufficient to draw a conclusion. A hundred? Once again, if we only know about descriptive statistics this is just a matter of taste. Secondly, has anyone considered how a hundred measurements would affect Sven's marriage? This may sound like a humorous remark but the fact is that it costs both time and money to collect data. It is necessary at some point to decide how much effort it is reasonable to invest in our measurements.

The aim of these anecdotes is not to amuse ourselves at the expense of other people but to point out how difficult it can be to draw conclusions using descriptive statistics. We are now starting to realize that we need a different approach to solve this problem.

Just as in the tea experiment at the beginning of this chapter, we will start by considering what we can say about the distribution from which Sven's measurements were sampled. Firstly, we need to decide what we mean by the "real" fuel consumption of the car. This number is affected by a wealth of factors, including Sven's driving style and his typical drive cycle. The sample can tell us something about what the typical fuel consumption is when *he* is driving the car, *where* he is driving it. It will not tell us much about the fuel consumption when others are driving it elsewhere. The car itself does not have a "true" fuel consumption, since the driver is such an important part of the equation. Let's rephrase our question as follows: is the mean of the *population* from which Sven's fuel consumptions are sampled less than 6.0 liters per 100 kilometers? Since the variation in the fuel consumption is affected by many factors we will resort to the central limit effect and assume that it is normally distributed. Since our sample is small we may use the *t*-distribution. The mean $\bar{x}$ and standard deviation s of our sample then represent an observed *t*-value:

$$t_{obs} = \frac{\bar{x} - \mu}{s/\sqrt{n}} \tag{8.1}$$

Putting our $s = 0.1643$, $\bar{x} = 5.98$, and $n = 5$ into Equation 8.1 and letting the hypothetical mean μ be equal to 6.0, we obtain $t_{obs} = -0.27$.

The logic behind using the *t*-distribution is that the sample mean is expected to display some amount of random variation about the distribution mean, determined by the number of degrees of freedom. We use 6.0 liters per 100 km as a hypothetical distribution mean, since the claim is that the sample is drawn from a distribution with a mean *different* from this. The null hypothesis assumes that any deviation from the mean is due to random variation about the hypothetical mean. The significance test will determine whether this assumption may be rejected or not.

By rephrasing our question we have almost formulated our null hypothesis. Remember that we cannot prove that the fuel consumption is less than 6.0 liters per 100 km, but we may reject the statement that it is not based on a significance test. We will set up the following two hypotheses:

$$\begin{cases} H_0 : \mu \geq 6.0 \\ H_a : \mu < 6.0 \end{cases} \tag{8.2}$$

H_0 is the null hypothesis and assumes that the claim is incorrect – that the distribution mean is equal to 6.0, or even higher. (As we are making a one-sided test, variation about a value *higher* than 6.0 will support the null hypothesis also). H_a is called the alternative hypothesis and it represents the effect that we want to demonstrate by giving the data a chance to reject the null hypothesis.

To make a significance test we need a critical *t*-value. If we want to test the null hypothesis at the 95% confidence level the probability of observing a *t*-value more extreme than the critical value should be $\alpha = 0.05$. Looking in the table of probability points for the *t*-distribution in the Appendix, we find that $t_{crit} = 2.132$ for $\alpha = 0.05$ and four degrees of freedom.

Two things should be noted. Firstly, since the consumption is claimed to be *below* a certain value we are assessing a one-sided interval. As there is only one way of being outside a one-sided interval, we are not using $\alpha/2$ as we did for the two-sided confidence

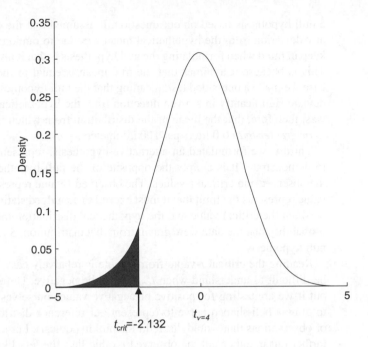

Figure 8.2
The *t*-distribution connected with Sven's fuel consumption measurements in the summer.

interval for blood pressures at the end of Chapter 7. Secondly, the table only provides positive *t*-values and we are testing for negative values. This is because we would reject the null hypothesis if our data were to yield a t_{obs} at the far left of the distribution in Figure 8.2. However, since the *t*-distribution is symmetric about the mean we may simply mirror it through the origin to obtain the relevant t_{crit} of -2.132, which is marked in Figure 8.2.

We note that our t_{obs} of -0.27 is less extreme than our t_{crit} of -2.132. This means that the data support the null hypothesis: the mean fuel consumption of Sven's car cannot be claimed to be less than 6.0 liters per 100 kilometers. This should come as no surprise, as you calculated a confidence interval for the mean fuel consumption in Exercise 6.9 that extended from 5.8 to 6.2 liters per 100 kilometers.

> **Exercise 8.1:** Calculate what mean fuel consumption is needed to obtain a statistically significant result. Use the same value of the sample standard deviation as above, $s = 0.1643$.

> **Exercise 8.2:** Calculate the one-sided confidence interval for the summer fuel consumption of Sven's car. (Recall that you calculated a two-sided interval in Exercise 6.9.)

The analysis that we just went through is called a one-sample *t*-test. It is used when we want to compare a sample to a fixed standard, such as a target fuel consumption. Let us review how we did it. Firstly, we stated our problem as a practical question. Is the mean fuel consumption really less than 6.0 liters per 100 kilometers? Secondly, we formulated a null hypothesis based on our question. It assumed that the claim was incorrect and that any deviation from the hypothetical mean was due to random variation. A general rule to keep in mind when formulating the null hypothesis is that it always includes an equals sign. This is because it assumes that the true mean is equal to the hypothetical mean. In this case we made a one-sided test, meaning that the null hypothesis is rejected only if the data deviate significantly in a given direction from the hypothetical mean. Our null hypothesis was, therefore, that the mean of the distribution from which the data were drawn is *equal to or greater than* 6.0 liters per 100 kilometers.

Thirdly, we formulated an alternative hypothesis, representing the claim that we want to demonstrate. It is simply the opposite of the null hypothesis. Finally, we determined the observed and critical *t*-values. The observed *t*-value represents the data and the critical value represents the limit that it must exceed to be judged statistically significant. If t_{obs} lies between the critical value and the hypothetical distribution mean, there is more than 95% probability that the data were drawn from that distribution. Such a result would support the null hypothesis.

Reading the critical *t*-value from a table is relatively easy but, as we have seen, it can be difficult to understand when t_{crit} should be negative. Unfortunately, we have to figure out if we are testing for positive or negative values ourselves. The easiest way to do it is to draw a bell-shaped curve to represent our reference distribution and visualize the sort of observations that would discredit the null hypothesis. For a two-sided test it is easy: the farther out in either tail the observed *t*-value lies, the less likely the null hypothesis is to be true. In one-sided tests it is necessary to understand if observations to the right or to the left of the mean are less likely under the null hypothesis. t_{crit} is positive if observations to the right would discredit the null hypothesis, otherwise it is negative. These scenarios are summarized in Table 8.2.

When hypothesis tests are performed using statistical software the result is often presented as a so-called *p*-value. This value is just the probability that the sample in question was drawn from the distribution associated with the null hypothesis. The lower this probability is, the less credible the null hypothesis becomes. The *p*-value of t_{crit} is, by definition, our α-value (usually 0.05). If t_{obs} is more extreme than t_{crit}, *p* will be lower than α. This outcome supports the alternative hypothesis and the result is said to be *statistically significant*. Interpreting the *p*-value is often confusing to beginners, as it may seem counter-intuitive that a low probability represents a statistically significant result. As researchers we are generally more interested in the alternative hypothesis. It is important to remember that the *p*-value is the probability of obtaining data that are at least as extreme as the ones in the sample, assuming that the *null hypothesis* is correct. If you are in doubt you may find the following mnemonic helpful:

If *p* is low, H_0 should go!

Before leaving the one-sample *t*-test it is appropriate to mention that our use of the *t*-distribution relies on the assumption that the data are reasonably normally distributed.

Table 8.2 When to reject the null hypothesis.

Null hypothesis	Type of test	Reject H_0 if:
$\mu \leq \mu_0$	One-sided	$t_{obs} > t_{crit}$
$\mu \geq \mu_0$	One-sided	$t_{obs} < -t_{crit}$
$\mu = \mu_0$	Two-sided	$t_{obs} < -t_{crit}$ or $t_{obs} > t_{crit}$

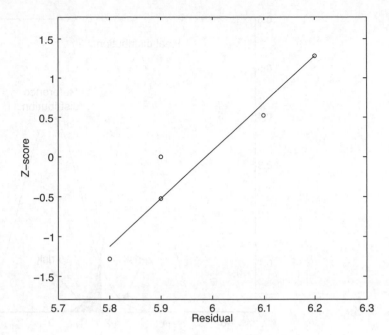

Figure 8.3
Normal plot for the fuel consumption data.

This is often a fair assumption but it is best to convince oneself before continuing the test. The easiest way to do it is to make a normal plot of the data since it may be difficult to judge the normality from a histogram when the sample is small. It is a general recommendation to always plot your data before analyzing them, as a graphical representation provides a better overview over the distribution and relationships in the data than the numbers themselves.

You should not expect all the points to adhere closely to a straight line in the normal plot but you should look out for obvious signs that there is skewing. Even severely skewed samples may be useful, as skewing often can be remedied by a data transformation, such as taking the logarithm of the data or something similar. It is also good to know that moderate violations of normality are not a concern. All hypothesis tests presented in this chapter are robust to non-normality, especially when the sample size increases. Figure 8.3 shows a normal plot of the fuel consumption data. This sample is too small to allow an accurate assessment of the normality but at least there are no apparent signs of skewing.

Exercise 8.3: A brand of breakfast cereals states that the package contains 750 g. You weigh the contents of six packages and get the following weights:

| 748 | 759 | 756 | 753 | 750 | 764 |

You suspect that the true mean weight could be less than 750 g. Test if this is the case at the 95% confidence level. (The null hypothesis is H_0: $\mu \geq 750$ g.)

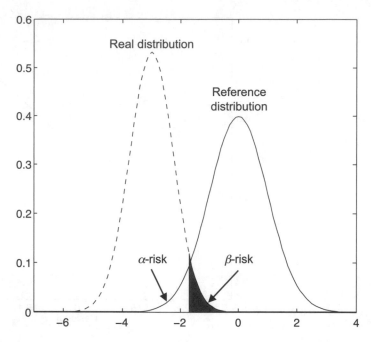

Figure 8.4
Illustration of how the β-risk occurs. When sampling from the left distribution, some tail values appear to support the null hypothesis although it is false. The null hypothesis is connected with the right distribution.

8.5   The Power of a Test

As we have seen, it is possible that extreme values occur purely by chance. This means that there is a small probability that our significance test rejects the null hypothesis even when it is true. This is called a Type I error and the probability of making one is, of course, equal to our significance level, α. This may be illustrated using Figure 7.15, which represents the reference distribution connected with a null hypothesis. It is perfectly possible to obtain values in the shaded tails of this distribution also when the null hypothesis is true, but our significance test will reject the null hypothesis when values that occur with a probability of less than $\alpha = 0.05$ are drawn.

When designing an experiment it is also important to be aware of a different risk: that of accepting the null hypothesis when it is, in fact, false. This is called a Type II error and the probability of making one is represented by the symbol β. Figure 8.4 illustrates how this risk occurs. The solid curve on the right is the reference distribution associated with the null hypothesis. It could, for example, be the t-distribution connected with Sven's fuel consumptions if the true mean were 6.0 liters per 100 km. Let us say that the null hypothesis is false. This means that we are sampling from another population, which is represented by the left, dashed curve. In actual situations we never know exactly which distribution we are sampling from, so it is only drawn here for illustration. The significance test tells us nothing of the real distribution – it only tells us whether it is probable that we are sampling from the reference distribution or not.

		Decision	
		Accept H_0	Reject H_0
Actual situation	H_0 is true	$1-\alpha$ *(Confidence level)*	α Type I error *(Significance level)*
	H_0 is false	β Type II error	$1-\beta$ *(Power of test)*

Figure 8.5
α is the probability of a Type I error, β the probability of a Type II error. These probabilities define the power, significance level and confidence level of the test.

What complicates the situation in Figure 8.4 is that the tails of the two distributions overlap. We may therefore obtain a value from the real distribution that *could* have been drawn from the reference distribution with a probability greater than α. The null hypothesis would then appear to be supported by the data although it is false. A shaded area in the tail of the left distribution in Figure 8.4 represents the probability β of obtaining such a value from the left distribution. This probability is also known as the β-risk. The white area under the tail of the reference distribution represents α.

In practical terms, this means that there is always a certain probability that a real effect remains undetected in your experiment. When you design your experiment you must decide how large a β you are willing to accept. What should the probability be of letting the effect go undetected? Zero is not an option.

Figure 8.4 shows us that the β-risk depends on the overlap between the two distributions. This, in turn, depends on the significance level α, the difference δ between the distribution means, and their respective widths. According to the central limit theorem the width decreases with increasing sample size. This means that the smaller δ is, the larger a sample you will need to detect it.

Figure 8.5 summarizes the four categories of possible outcomes of an experiment. On the left we have the two possibilities for the null hypothesis: it is either true or false. The two decisions that are possible to make based on the data are given at the top: the null hypothesis is either accepted or rejected. As we have mentioned, two of the outcomes are incorrect. A Type I error, rejecting the null hypothesis when it is true, occurs with probability α. A Type II error, accepting the null hypothesis when it is false, occurs with probability β. This leaves two categories of correct outcomes. Accepting the null hypothesis when it is true occurs with probability $1 - \alpha$, which we refer to as the *confidence level* of our test. The last category, rejecting the null hypothesis when it is false, occurs with probability $1 - \beta$. This number is called the *power* of the test. Just as 95% is a conventional confidence level in many experiments, 80% is a typical power, but both these numbers may vary according to the situation. In applications like medicine and aerospace engineering, where it is highly important to avoid errors, higher confidence levels and powers are often used.

If you find it difficult to remember these categories of error, it might be useful to compare an experiment to a court of law. For the prosecutor, the purpose of a trial is to give the evidence a chance to reject the "null hypothesis" that the suspect is innocent. A Type I error then corresponds to convicting a suspect that is, in fact, innocent. Freeing a guilty person would be a Type II error. Just as in experiments, one way for the prosecutor to minimize this risk is to gather more evidence against the suspect.

The problem that remains is to determine how large a sample we need to detect a certain difference, δ, between two population means at a certain level of power. Most statistical software contains a function that does this, which is often called "power and sample size" or something similar. It typically requires that we specify four of the five parameters α, $1 - \beta$, δ, the standard deviation, and the sample size. Based on these it calculates the fifth for us.

The standard deviation describes how far the estimated mean deviates from the true mean. This is not known in practice, so the power analysis can be seen as a play with probabilities. It shows what would happen if the true standard deviation were the one that we specify. Let's use Sven's fuel consumptions as an example. The sample size is five. The standard deviation of the mean is estimated by the sample standard deviation divided by the square root of five. Say that we are interested in using a significance level, α, of 0.05 and a power, $1-\beta$, of 0.8. Plugging these values into the power analysis of a statistical software package returns the fifth parameter, $\delta = 0.12$. This means that, with the specified standard deviation and sample size, the true mean has to deviate by more than 0.12 from the hypothetical mean (6.0 liters per 100 km) to be considered different from it. In Sven's data the difference, δ, between the sample mean and the hypothetical mean is only 0.02 liters per 100 km. How large a sample would be needed to detect this difference? Using the same standard deviation, α, and β as before, the software returns a sample size of 108. Sven's data set is indeed insufficient to reveal a difference this small.

8.6 Comparing Two Samples

The one-sample t-test is used to compare one sample to a target value. It is common, for instance, in engineering when a product is developed to meet a certain specification limit. In science, it is more common to compare two samples to each other. Scientists are often interested in assessing the existence of an effect of an environmental condition, a medical procedure, an experimental treatment, or something similar. They do this by comparing samples drawn under different conditions. In such cases there are two scenarios that can occur, since two samples can either be *independent* of or *dependent* on each other. If we consider a medical study, we could assign one group of people to a certain treatment and use another group as a control group. This would produce two independent samples. If we instead test the same group before and after receiving the treatment we get two dependent samples, since every person is associated with an observation in each sample. In dependent samples an observation in one sample can be paired with a corresponding observation in the other.

When comparing two *independent* samples to each other, we use the so-called two-sample t-test. To familiarize ourselves with it we will analyze an experiment where the yield of a chemical synthesis is tested using two different solvents. To maintain the integrity of the significance test the solvent is chosen randomly before each run, in a manner that results in 10 observations for each solvent. The data are given in Table 8.3 and plotted in Figure 8.6. We see that, although sample one has a lower mean yield, there is an overlap between the samples and substantial variation within each set. Let us see if the difference between them is statistically significant.

The two-sample t-test requires that the samples have equal variances. Say that, based on our experience of similar experiments, we find this to be a reasonable assumption. In

Table 8.3 Data from the synthesis experiment.

	Sample 1	Sample 2
	85.2	87.5
	84.5	86.7
	82.5	85.7
	83.9	87.9
	83.6	86.9
	82.7	84.8
	85.5	84.2
	83.8	87.5
	83.9	86.4
	86.4	83.8
Mean:	84.2	86.14

such a case it is adequate to estimate the variance by pooling the sums of squares of the two samples, since a larger sample size provides a better estimate. The variance of the first sample is simply its sum of squares divided by its number of degrees of freedom:

$$s_1^2 = \frac{SS_1}{n_1 - 1} \tag{8.3}$$

and the variance of the second sample is determined in the same way. If these two variances are reasonably equal they may be combined to provide a pooled estimate of the variance:

$$s_{pooled}^2 = \frac{SS_1 + SS_2}{n_1 + n_2 - 2} \tag{8.4}$$

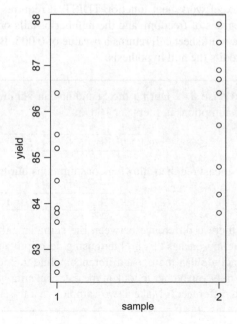

Figure 8.6
Plot of the data from the synthesis experiment.

Here, we have simply added the sums of squares in the nominator and the numbers of degrees of freedom in the denominator.

We want to determine if the difference between the means of the two samples is significantly different from a hypothetical difference δ. Our observed t-value then becomes:

$$t_{obs} = \frac{(\bar{y}_1 - \bar{y}_2) - \delta}{s_{pooled}\sqrt{\frac{1}{n_1} + \frac{1}{n_2}}} \tag{8.5}$$

where the denominator is the estimated standard deviation of the difference between the means. (You may convince yourself of this by using the additive property of the variances of the two means.) Comparing with Equation 8.1 you will find that Equation 8.5 is exactly analogous but instead of sample and population means we are now using *differences* between sample and population means.

The null hypothesis assumes that the two solvents produce equal results and that any difference between them is due to random error, so:

$$\begin{cases} H_0 : \delta = 0 \\ H_a : \delta \neq 0 \end{cases} \tag{8.6}$$

Note that the null hypothesis, as always, contains an equals sign.

Using $\delta = 0$ in Equation 8.5 we obtain a t_{obs} of -3.16. To see if this is a statistically significant result we need a critical t-value. Just as before, it may be read from the table of the t-distribution in the Appendix. With 18 degrees of freedom and $\alpha/2 = 0.025$ (that is, for a two-sided test) we obtain a t_{crit} of 2.101. Since t_{obs} is farther from zero than ± 2.101 we conclude that the result is significant – the data support the conclusion that there is a difference between the samples.

An alternative way of making the significance test is to calculate the p-value directly, using the Excel worksheet function TDIST. It requires three input arguments: t_{obs}, the number of degrees of freedom, and the number of tails of the test. Writing "=TDIST(3.16;18;2)" into a worksheet cell returns a p-value of 0.005. Being much lower than our α of 0.05, this discredits the null hypothesis.

Exercise 8.4: During the Scandinavian summer, Sven measured the following fuel consumptions in liters per 100 km:

5.9	5.8	5.9	6.2	6.1

He measured the following consumptions during the winter:

5.9	6.3	6.0	6.1	6.0

Is there a difference between the summer and winter fuel consumption? In a class that approaches this problem using descriptive statistics, half of the students typically conclude that there is a difference, while the other half holds that there is not. (Some do not even bother to look at the data, but simply refer to the well-known fact that there *is* a difference!) Make a two-sample t-test to settle the question.

Let us look at another example in order to demonstrate how dangerous it can be to follow cookbook recipes without reflecting over the nature of the data. Recall the team

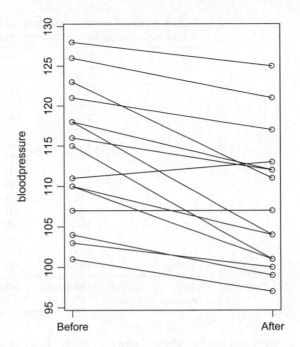

Figure 8.7
Blood pressure data.

of researchers who, in Chapter 7, wanted to investigate how a certain diet affected the blood pressure of healthy people. They measured the systolic blood pressure of their 15 test persons before starting the diet. The values, in mm Hg, are given for persons 1–15 below. They are rounded to the nearest integer to help us concentrate on the essentials:

128	126	123	121	118	118	116	115	111	110
110		107		104		103		101	

After 15 days of diet, the blood pressures were measured again:

125	121	111	117	112	104	112	101	113	104
101		107		99		100		97	

Again, these values are ordered from test person 1 to test person 15. Before continuing the analysis we examine them graphically in Figure 8.7, which shows a clear tendency for the blood pressure to drop after the diet. Only one person shows an increase and another person's remains the same after the test.

We will now attempt to investigate the significance of this tendency by the two-sample t-test. The null hypothesis is that there is no difference between the samples, or that $\delta = 0$. The mean blood pressures before and after the diet are 114.1 and 108.3 mm Hg, respectively. Using these values in Equation 8.5 we obtain $t_{obs} = -1.89$. We read the critical t-value from the table of the t-distribution in the Appendix, which states that $t_{crit} = 2.048$ for $\alpha/2 = 0.025$ (two-sided test) and 28 degrees of freedom. As t_{obs} is less extreme than t_{crit} the null hypothesis is supported: there seems to be no difference between the samples.

Table 8.4 Blood pressure data recorded before and after the diet, as well as the difference between them.

	Before	After	Difference
	128	125	−3
	126	121	−5
	123	111	−12
	121	117	−4
	118	112	−6
	118	104	−14
	116	112	−4
	115	101	−14
	111	113	2
	110	104	−6
	110	101	−9
	107	107	0
	104	99	−5
	103	100	−3
	101	97	−4
Mean:	114.1	108.3	−5.8
Standard deviation:	8.4	8.5	4.7

How can this be, when Figure 8.7 clearly shows that the blood pressure decreases after the diet? The reason for this counter-intuitive result is simply that we have used the wrong test for these data. Since the same persons are associated with one observation in each sample, the samples are *dependent*. In such cases we should use what is called the *paired* *t*-test, which will be explained in the remainder of this section. It will also become clear why the paired *t*-test works better in this case.

All the theory needed to understand the paired *t*-test has already been covered, since it is simply a special case of the one-sample *t*-test. The first step is to calculate the change in blood pressure that each test person experiences as a result of the diet, just as we did at the end of Chapter 7. These are given in the third column of Table 8.4. Looking at the standard deviations of the columns, which are given at the bottom of the table, we see that the "difference" column displays less variation than the original data in the "before" and "after" columns. The reason is that the original data contains variation between the *individuals*. Taking the difference between each person's value before and after the diet we eliminate this source of variation. What is left is each person's response to the diet, irrespective of his or her original blood pressure. In short, the paired *t*-test removes variation that is not due to the experimental treatment and highlights the effect of the treatment.

The difference column in Table 8.4 can now be treated as a single sample. We obtain the observed *t*-value using Equation 8.1, where $\bar{x}$ now corresponds to the mean difference and μ to the hypothetical difference, which is zero. The mean difference in blood pressure is −5.8 mm Hg, the sample standard deviation is 4.7 mm Hg, and the number of observations is 15. Plugging these numbers into Equation 8.1 yields a t_{obs} of −4.62, which is much farther from the hypothesized mean of zero than our t_{crit} which, according to the *t*-table in the Appendix, is 2.145. Alternatively, the *p*-value returned from the worksheet function TDIST is 0.0004. As opposed to the result of our erroneous use of the two-sample *t*-test with these data, this analysis is capable of detecting the effect that the diet has on the blood pressure. The reason that the paired *t*-test is more sensitive is that it removes the variation between the individuals, but this is only possible when the samples are dependent.

8.7 Analysis of Variance (ANOVA)

So far we have compared one sample to a standard (the one-sample *t*-test) and two samples to each other (the two-sample and paired *t*-tests). Sometimes we want to find out if there is a difference between several samples. A technique that is useful in such cases is the so-called Analysis of Variance, or ANOVA for short. It is based on breaking the variation in the data down into several parts.

It may seem confusing at first that ANOVA, which is a technique for comparing three or more sample means, is based on the variation in the data. The best way to understand how this works is probably to work through an example. We will compare three different surface treatment methods for steel. An experiment is performed where steel parts are surface treated and subjected to abrasive wear tests. After the tests the weight loss is measured. We are interested in finding out if the surface treatments affect the amount of wear. The measurement data are given in Table 8.5, where the columns represent the treatments (A, B and C) and the rows correspond to five different metal parts exposed to each treatment. In total, 15 parts are used in this experiment. Below the table, the mean weight loss is given for each treatment and for the experiment as a whole (grand mean). Also, the differences between the grand mean and the treatment means are given. It is important to point out that the metal parts are allocated randomly to the treatments and tested in random order, to ensure the integrity of the significance test that we are about to perform.

It cannot be stated often enough that it is always appropriate to plot the data before proceeding further with the analysis. The three samples are shown in Figure 8.8, where the horizontal line represents the grand mean. The first five points represent sample A, the next five points sample B, and the last points sample C. Each point is connected to the grand mean with a vertical line that represents the deviation from the grand mean. It is quite easy to see that the samples have different means, but at what level of significance? There is substantial variation within each treatment and the samples are partially overlapping. Figure 8.9 shows the same data but the points are now connected to their sample means, represented by the three horizontal lines. We can now make the important observation that the average deviations from the sample means are smaller than the average deviation from the grand mean. We will return to this observation after going through the analysis.

To determine the total variation in the data we will calculate the deviations from the grand mean. These are given in the matrix **D** in Table 8.6 and are just each element in Table 8.5 minus the grand mean. The first row contains $-2.7 = 11.3 - 14$, $4.7 = 18.7 - 14$, and so on. The total variation is then decomposed into a part **T** due to the treatments

Table 8.5 Measurement data from abrasive wear tests in the surface treatment experiment.

	A	B	C
	11.3	18.7	12.0
	12.5	17.3	9.8
	18.5	16.7	8.2
	11.9	16.6	14.6
	15.2	16.9	10.3
mean:	13.9	17.2	11.0
grand mean:		**14.0**	
difference:	−0.1	3.2	−3.0

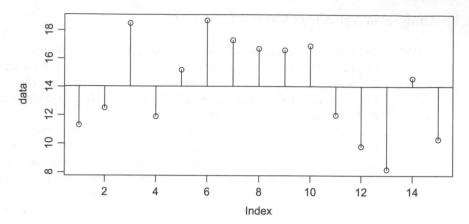

Figure 8.8
Abrasive wear data plotted in relation to the grand mean. The first five points represent sample A, the next five sample B, and the last five sample C.

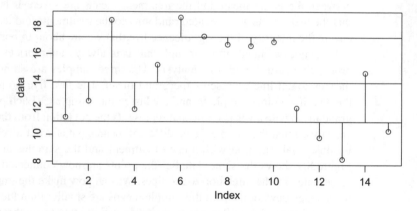

Figure 8.9
Abrasive wear data plotted in relation to the respective sample means.

and a residual part **R**. **T** contains the treatment means minus the grand mean. It represents the variation *between* the treatments. Since **R** represents the variation that is *not* due to the treatments, it contains the elements of Table 8.5 minus the treatment means. This is the variation *within* the treatments or, in other words, the experimental error. Note that if you add the elements of **T** and **R** you obtain the elements in **D**.

Table 8.6 Abrasive wear data broken down into several parts.

	D		=		T		+		R	
−2.7	4.7	−2.0		−0.1	3.2	−3.0		−2.6	1.5	1.0
−1.5	3.3	−4.2		−0.1	3.2	−3.0		−1.4	0.1	−1.2
4.5	2.7	−5.8		−0.1	3.2	−3.0		4.6	−0.5	−2.8
−2.1	2.6	0.6		−0.1	3.2	−3.0		−2.0	−0.6	3.6
1.2	2.9	−3.7		−0.1	3.2	−3.0		1.3	−0.3	−0.7

Table 8.7 ANOVA table for the abrasive wear tests.

SOURCE	ν	SS	MS	F	P
Treatment	2	98.16	49.08	9.46	0.003
Error	12	62.25	5.19		
Total	14	160.41			

The data in Table 8.6 will now be used to construct what is usually called an ANOVA table, given in Table 8.7. The variation is divided into two sources: that due to the treatments (variation *between* samples) and that due to experimental error (variation *within* samples). The last source is just the sum of the others, or the *total* variation. We will start by looking at the second column, which contains the sums of squares, SS, from the three sources. SS_R is due to the experimental error and is calculated by simply squaring all the elements of **R** and summing them:

$$SS_R = (-2.6)^2 + (1.5)^2 + \cdots + (-0.7)^2 = 62.25 \tag{8.7}$$

Similarly, SS_T, which is due to the treatments, is all the elements of **T** squared and summed. The total sum of squares, SS_D, is the elements of **D** squared and summed. You may note that it is also given by $SS_T + SS_R$, as the sums of squares are additive.

The first column of Table 8.7 contains the numbers of degrees of freedom for each source of variation. Recall from Chapter 7 that they are given by the number of observations and the number of population parameters that are estimated from them. The total variation is associated with the matrix **D**. Since it is the difference between each observation and the grand mean it has $n - 1 = 14$ degrees of freedom. The error is associated with the matrix **R**, where treatment means have been calculated from the five observations in each column. As there are three treatment means, it has $15 - 3 = 12$ degrees of freedom. Finally, **T**, which contains the differences between the treatment means and the grand mean, represents the variation due to the treatments. There are three treatments, so it has $3 - 1 = 2$ degrees of freedom. Note that the total number of degrees of freedom is the sum of the others.

We now have all the elements needed to calculate the mean squares, MS, in the third column. These are just each sum of squares, SS, divided by its number of degrees of freedom, ν. They therefore have the form of variances. The error mean square is actually the unbiased estimate of the error variance, σ^2, that was introduced in Box 6.1. It therefore says something about the precision in the data. If the treatments do *not* produce different results (that is, if the treatment means are equal), it can be shown also that the treatment mean square is an unbiased estimate of σ^2 [2]. In this case, Table 8.7 shows that the treatments produce a larger mean square than the error does. Comparing these two components of the variation thereby reveals that there is a difference between the treatment means. The question is if it is statistically significant.

To answer this question we need a suitable significance test. The null hypothesis to be tested is that the treatments do *not* have an effect, which is equivalent to saying that both the treatment and error mean squares are estimates of σ^2. The equality of two variances is tested using the F-test, since ratios of variances can be shown to follow a so-called F-distribution. The interested reader can find out why this is so by reading, for example, Box, Hunter, and Hunter [3]. Here we will settle with just stating this as a fact. The F-test follows the same basic procedure as the t-test. We will compare an observed F-value, representing our data, to a critical F-value, representing our α-value. This will indicate the potential effect of the treatment.

To get our observed F-value of 9.46, we simply divide the mean square of the treatment by that of the error. This is to be compared to the critical F-value for the 95% confidence level, which can be read from a table of the F-distribution. The procedure is exactly analogous to the t-test: if our observed F-value is more extreme than the critical one, the null hypothesis is discarded. In this example, however, we will proceed directly to calculating a p-value as this is so easily done using the Excel function FDIST.

The required input arguments are the observed F-value and the number of degrees of freedom of the two mean squares. Thus, writing "=FDIST(9.46;2;12)" into an Excel worksheet and pressing enter returns the p-value, 0.003. This is the probability of obtaining our data, or something more extreme, under the null hypothesis that the treatments have no effect on the result. The outcome is clearly significant at the 95% confidence level. The null hypothesis is thereby strongly discredited. Our data support that that the surface treatments produce different amounts of weight loss. Remember the mnemonic: "if p is low, H_0 should go!'

In summary, we have used the ANOVA to judge the significance of the difference between the sums of squares computed from the grand mean and from the individual treatment means. To put it simply, if the means are significantly different, the treatment sum of squares is smaller than the overall sum of squares [4]. This is actually what we saw graphically in Figures 8.8 and 8.9 when we noted that the average deviation from the sample means were smaller than the average deviation from the grand mean.

The ANOVA is used to compare three or more sample means to each other. It is worth mentioning that if you were to try to compare two sample means by ANOVA you would get exactly the same result as from a two-sample t-test. In fact, it is a good exercise to convince yourself of this by doing it.

It may also be worth mentioning that, if you try to reconstruct the entries in the ANOVA table using the numbers in Tables 8.4 and 8.5, you will get a slightly different result as they have been rounded to one decimal point.

Exercise 8.5: Our gardener friend from Chapter 2, Mr. Green, gets hooked on the idea of growing giant sunflowers. He wonders if he can affect the size of the flower heads by adding fertilizer and decides to try three soil treatments. The first is organic soil, based on compost from his own garden. The second uses the same soil but with addition of a high nitrogen formula throughout the growing season. He suspects that phosphorus may affect the growth of the flower (remember that he has a Ph.D. in plant physiology) and decides, for the third treatment, to begin with the high nitrogen formula and change to a high phosphorus formula as soon as the flower heads begin to form. He realizes that it is important to randomize the treatments to avoid systematic errors due to local temperature variations, how well drained the soil is, and so on. To maintain control over the treatments he decides to plant seeds from a single parent flower in pots and attribute the treatments to them randomly, by rolling a die. He places the pots in random order along the south wall of his house. 80 days after sowing he measures the flower head diameters (Table 8.8). The measures are rounded to the nearest quarter inch.

Use an ANOVA to find out if the difference between Mr. Green's experimental treatments is statistically significant.

Table 8.8 Flower head diameters.

O	N	N+P
12.25	11.25	13.00
11.50	12.75	13.50
10.75	11.75	12.50
10.50	12.00	12.25

O: organic; N: nitrogen; N+P: nitrogen and phosphorus

8.8   A Measurement System Analysis

The ANOVA is a versatile tool that can be used in many situations where several influences could affect a response. In this section we are going to use a variant called *two-way* ANOVA to analyze a measurement system. It is used when the variation in the samples under study occurs under the influence of two variables, or *factors*.

An engine laboratory wants to determine the precision of its measurement system for diesel engines. It measures the exhaust emissions of particulates using a so-called smoke meter. These emissions are of interest since the laboratory has found that the smoke value is sensitive to many parameters and, therefore, the repeatability of these measurements is typically poorer than for other emissions.

The laboratory has, wisely, adapted the routine to run a stability check every morning, which consists of taking measurements while operating the engine at a standard condition. If the equipment is working correctly there should be no drift in these measurements from day to day. There should only be the natural variation about a mean value that is always expected and which represents the precision of the measurements. (The accuracy is, of course, taken care of by calibrating the instruments.) Looking in its databases it finds a series of 24 stability checks that were made using the same engine. This sample size should be adequate for determining the measurement precision.

Two potential shortcomings of the data set are found. Firstly, two different operators have collected the data. This could introduce systematic variation if the operators run the tests in slightly different ways. Secondly, due to an electronic failure, the engine's electronic control unit (ECU) was replaced halfway through the series. Although it is unlikely, it cannot be ruled out that this, too, could introduce systematic variation in the data.

The laboratory engineers decide to use ANOVA to investigate if the operator has an effect. If the ECU was to affect the smoke level this, too, would introduce variability in the data. As it would not be appropriate to include this variation in the measurement precision, they decide to treat the ECU as a second factor in a two-way ANOVA. We will see how this is done shortly but first it is appropriate to mention a third drawback of the data set. We have previously discussed the necessity of randomizing the order of the treatments in an experiment. With these historical data we cannot attend directly to this need, as we have no control over the data collection process. In this case, the order in which the operators had obtained the data did not show any apparent patterns. On the other hand, all measurements that were made with the second ECU were, of course, made after the first one broke. This means that, if the ECU were to cause an effect, we would not be able to determine if it were due to the ECU or to some other change that happened to occur while the ECU was exchanged. Keeping this in mind, it was still judged appropriate to assess the potential effect of the ECU.

Table 8.9   Data for the measurement system analysis.

| | OPERATOR | | |
ECU	Clark	Lois	ECU mean:
1	0.59	0.57	
1	0.55	0.49	
1	0.52	0.48	0.52
1	0.51	0.47	
1	0.51	0.50	
1	0.54	0.49	
2	0.48	0.45	
2	0.54	0.51	
2	0.54	0.49	0.49
2	0.55	0.52	
2	0.45	0.44	
2	0.50	0.45	
Operator mean:	0.52	0.49	
Grand mean:	**0.51**		

It should be noted that the measurement system includes more than the smoke meter. It includes all parts that affect the measurement result, including the engine, its control system and the operator. By breaking the variation down into several parts we may get a better idea of how the measurement precision could be improved.

The 24 observations are displayed in Table 8.9, where the columns represent the two operators, Clark and Lois. The rows are divided into two groups according to the ECU that was used during the measurements. It also shows the operator means at the bottom and the ECU means on the right, as well as the grand mean. The data are plotted in Figure 8.10,

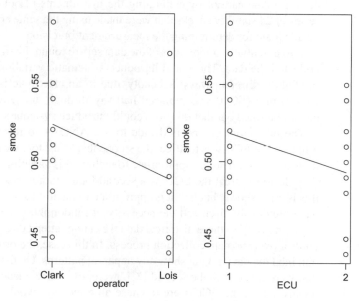

Figure 8.10
Measurement system analysis data.

Table 8.10 Measurement system data broken down into several parts.

D		=	O		+	E		+	R	
0.09	0.06		0.02	−0.02		0.01	0.01		0.06	0.07
0.04	−0.02		0.02	−0.02		0.01	0.01		0.01	−0.01
0.01	−0.03		0.02	−0.02		0.01	0.01		−0.02	−0.02
0.00	−0.04		0.02	−0.02		0.01	0.01		−0.03	−0.03
0.01	−0.01		0.02	−0.02		0.01	0.01		−0.02	0.00
0.03	−0.02		0.02	−0.02		0.01	0.01		0.00	−0.01
−0.02	−0.06		0.02	−0.02		−0.01	−0.01		−0.03	−0.03
0.03	0.00		0.02	−0.02		−0.01	−0.01		0.03	0.03
0.03	−0.02		0.02	−0.02		−0.01	−0.01		0.03	0.01
0.04	0.01		0.02	−0.02		−0.01	−0.01		0.04	0.04
−0.05	−0.07		0.02	−0.02		−0.01	−0.01		−0.06	−0.04
−0.01	−0.06		0.02	−0.02		−0.01	−0.01		−0.01	−0.03
SS_D:	0.036		SS_O:	0.008		SS_E:	0.004		SS_R:	0.025

which shows a tendency toward lower smoke values in Lois' measurements and with ECU number 2.

As in the previous section we are going to determine the variation in the data and break it down into several parts. In Table 8.10, the matrix **D** represents the deviations from the grand mean of 0.51. It is obtained by subtracting the grand mean from every element in Table 8.9. The matrix **O** contains the variation due to the operator, obtained by subtracting the grand mean from the operator means. The matrix **E** contains the variation due to the ECU, or the difference between the ECU means and the grand mean. The remaining variation is represented by the residual matrix **R**, which is obtained by subtracting the elements of **O** and **E** from those in **D**. This is the part of the total variation that is neither explained by the operator nor by the ECU. Table 8.10 also displays the sums of squares, SS, for all of the sources of variation, obtained by squaring the elements of each matrix and adding them up.

Before we can fill out the ANOVA table we need to determine the numbers of degrees of freedom, v, for each of the sources of variation. The total variation, **D**, was determined by subtracting one estimated parameter (the grand mean) from the n observed values, so $v_D = n - 1$. Similarly, to obtain the variation due to the operator, **O**, we subtracted one parameter (the grand mean) from the two operator means, so $v_O = o - 1$, where o is the number of operators. **E** has $v_E = e - 1$ degrees of freedom, where e is the number of ECUs. Finally, the residual degrees of freedom are given by the total number of degrees of freedom minus those of the operators and the ECUs.

Exercise 8.6: Fill in the ANOVA entries for the measurement system analysis in Table 8.11. Follow the procedure that was used for the one-way ANOVA in the previous section and note that each of the factors (operators and ECUs) should be compared with the *measurement error*, determined by the residuals.

Make sure that you try to answer all of the following questions before proceeding further:

(a) Which factor has the largest effect on the smoke value?
(b) Which factor(s) has(ve) a *statistically significant* effect?
(c) Assuming that the smoke meter is in prime condition, think of a way to improve the precision of the measurement system.

Table 8.11 ANOVA table for the measurement system analysis – to be filled in by the reader.

SOURCE	v	SS	MS	F	p
Operator					
ECU					
Error					
Total					

If the entries in the ANOVA table are correct we should see that, although Figure 8.10 indicates that the ECU has an effect, it is not significant at the 95% confidence level ($p > 0.05$). By this criterion we would conclude that the effect is too small to stand out from the measurement error. On the other hand, the data strongly suggest that the operator has an effect on the measured smoke level ($p < 0.02$). It is the only factor that has a statistically significant effect. Let us discuss how we should interpret these results.

As mentioned in connection with the one-way ANOVA, the error mean square is an estimate of the error variance, σ^2. If neither the operator nor the ECU were to have an effect, this would be a measure of the measurement precision. A more practical measure of the precision would be the error standard deviation, as it has the same unit as the measured smoke values. It is estimated by the square root of the error mean square in Table 8.11, which is 0.035. This is the error bar that should be attached to the measured smoke values if the variation due to the operators could be eliminated.

In this case, however, the operators add to the variation in the data, decreasing the overall precision in the measurements. As previously mentioned, the measurement system includes all parts that affect the measured values, including the operator. Assuming that nothing can be done to decrease the error itself, our ANOVA suggests that the precision of the measurement system can be improved by doing something about the variation between the operators. How can this be done? Here are a few suggestions given by students:

- Increase the number of operators.
- Increase the number of measurements.
- Use one operator only.
- Fire Clark, or put him on a management track, and let Lois run the tests.

Again, these examples are not given to amuse ourselves at the expense of others, but to point out how difficult it can be to interpret a statistical analysis in practical terms before being used to thinking statistically. The first two suggestions are based on a misinterpretation of the central limit theorem, which states that the standard deviation of a *sample mean* is expected to decrease with increasing sample size. What we are interested in here is decreasing the standard deviation of a *sample*. If you are not sure about the difference between these two you should revisit the section about the central limit effect in Chapter 7.

The last two suggestions aim to increase the precision by eliminating the variation arising from using several operators. This will work, but they both miss an important point. The reason that Lois and Clark obtain different results from the stability check must be that they are running the tests *differently*. Before taking any steps it would be wise to find out where this difference arises and how it can be eliminated. In other words, the laboratory seems to need a standard procedure for operating the engine. The last suggestion is interesting, not only because it assumes that managers are poor engineers but because it assumes that

Lois must be *better* at running the tests since her smoke values are lower. It is, of course, important to decrease the exhaust emissions from the engine when it is used on the road. But if Lois obtains lower smoke emissions in the laboratory by running the engine warm for ten minutes longer than Clark, for instance, this represents no improvement under road conditions.

What the laboratory needs to do to improve the measurement precision in this case is simply to adopt a standard operating procedure to assure that there are no systematic differences between the operators.

8.9 Other Useful Hypothesis Tests

The hypothesis tests presented in this chapter are run-of-the-mill techniques that are used in experiments across many disciplines. So far, though, we have only treated techniques for analyzing the mean of one or several samples. Sometimes we are more interested in properties other than the mean, such as variances or proportions. Such analyses require different hypothesis tests. They will not be treated here but to make it easier for the reader to find the appropriate technique for a given situation, a few additional hypothesis tests and the situations in which they are used will be mentioned. Once we have understood the *t*-test it is quite straightforward to use other techniques, as all tests are based on the same principles:

- Firstly, formulate a question that contains a statistical parameter and a word expressing a difference. It could be "Is the sample mean different from the target value?", "Are the variances of these two samples different?", or something similar. In the teatime experiment it would be "Is the proportion of correct classifications higher than what could be expected to occur by chance?"
- Translate the question into a null hypothesis and an alternative hypothesis. These are mathematical relationships between the sample statistic and the population statistic. The null hypothesis assumes that the answer to the question is no – any apparent difference is due to natural, random variation in the data. As it assumes that no difference exists, the null hypothesis always contains an equals sign. The alternative hypothesis assumes that a real difference underlies the variation in the data – a difference that we would like the data to reveal. The alternative hypothesis is always the opposite of the null hypothesis. This means that, if the null hypothesis contains a "$=$", the alternative hypothesis contains a "$\neq$". If the null hypothesis contains a "$\leq$", the alternative hypothesis contains a "$>$", and so on.
- Find the reference distribution associated with the null hypothesis. This is the probability density function that the statistical parameter follows in the case that the process under study is governed completely by chance.
- Determine the critical value of the reference distribution at the desired level of confidence. Also determine the observed value of the test statistic, which is determined by your data. (This involves a variable transformation, for example from a sample mean to a *t*-value.) If the observed value is more extreme than the critical one the effect is deemed statistically significant, otherwise it is not.

The greatest difference between different hypothesis tests is the reference distribution that is used. At this point it should be clear that the reference distribution of the mean of a small sample is a *t*-distribution having the same number of degrees of freedom as the sample.

This is because the mean can be transformed into the variable t, according to Equation 8.1. Correspondingly, the *variance* of a sample can be transformed into a variable that follows what is called a χ^2-distribution – pronounced "chi-squared distribution". It is also characterized by the number of degrees of freedom of the sample. If we want to compare the variance of a sample to a target value we use a so-called χ^2-test for variance. If we instead want to compare the variances of two samples we use the so-called F-test. We have already been acquainted with the F-distribution in connection with the ANOVA, where we stated that ratios of variances follow this distribution. The F-distribution is characterized by the numbers of degrees of freedom of the two samples that are compared.

Some experiments yield results in the form of proportions. For example, if a team of engineers modifies a manufacturing process to improve product quality, it may be interested in knowing if the failure rate is lower among the items that have gone through the new process compared to those going through the original one. If you want to test proportions there are a number of techniques at your disposal, and the particular situation decides which one is better suited. Box, Hunter and Hunter [3] provide some examples of such tests.

8.10 Interpreting *p*-Values

It is common in many fields to report p-values from significance tests in scientific publications. Some statisticians are critical to this focus on the p-value because its meaning is frequently over- or misinterpreted. It is important to remember that the p-value is connected with the null hypothesis and nothing else. It measures the probability of obtaining data that are at least as extreme as the actual sample, under the assumption that the null hypothesis is true. Technically, this is not equivalent to the probability of the null hypothesis being true.

As explained in the example about Sven's fuel consumption, a significance test helps us to draw a more robust conclusion where a subjective judgment of the data could lead to ambiguities. Although the significance level is a somewhat arbitrary number, it is a quantitative means for motivating a conclusion. Nonetheless, the p-value is not entirely objective. Consider the t-test, for example. Here, the p-value is directly related to the observed t-value (a great t_{obs} corresponds to a small p). Equation 8.1 shows that the magnitude of t_{obs} depends on two things. It becomes large if the sample mean deviates substantially from the hypothesized population mean, that is, if there is a large effect in the data. But it also becomes large if the sample size n increases. In other words, it measures the effect in *relation* to the uncertainty in the data.

With a large sample size the uncertainty decreases and even small effects become statistically significant. Smaller samples, on the other hand, require the effect to be greater to stand out from the noise. In connection with the teatime experiment we expressed this by saying that a larger experiment is more sensitive than a small one. The same conclusion was reached in the section about the power of a test: to detect a small difference we need a large sample size. This means that the p-value says more about the precision in the data than about the reality of the investigated effect. A high p-value does not say that the effect is inexistent. It says that the experiment is too insensitive to detect it *if* it exists.

Due to this state of affairs the term "significant" is not entirely adequate, because even a small effect can produce a low p-value if the sample size is sufficiently large. We should at least say that an effect is *statistically* significant at a specified confidence level. Above all, we should remember that, to scientists, magnitudes of effects and experimental errors are of primary interest. We should not substitute these with a p-value but rather supplement them with one. After all, the *p*-value is a statement about your particular experiment more than a statement about reality.

8.11 Correlation

In research, as in life, most of the important questions have to do with relationships. An introductory treatment of statistical techniques for experiments would be lopsided without any mention of techniques for investigating relationships among variables. In the previous sections we have compared samples to each other. The reason why we compare them is, of course, that we expect them to differ from each other, probably because they were sampled under different conditions. In other words, we expect that the variables we sample are dependent on other variables. We expect the fuel consumption to vary with the season, the yield of a synthesis to vary with the solvent used, and the hardness of a material to depend on the surface treatment it has undergone. In the *t*-tests and the ANOVA the independent variables are *categorical*. This means that they can be labeled but not ordered – the seasons, solvents, or surface treatments have names but not numerical values.

The independent variable is frequently a numerical variable, such as a temperature, a price or a distance. In the remaining parts of this chapter we are going to see how relationships between numerical variables can be investigated. Firstly, we will focus our interest on the strength and direction of the relationship between two variables. Does one variable increase or decrease when the other increases? Such questions are answered using correlation analysis. In the next section, we will see how the relationship between the variables can be described mathematically, introducing a technique called regression modeling.

Correlation is a measure of the strength and direction of the linear association between two variables. Consider two samples of size *n* that were collected simultaneously, so that each observation in one sample is associated with an observation in the other. They could, for instance, consist of temperatures and pressures in the cylinder of an engine collected during compression. The correlation between two such samples is given by Pearson's correlation coefficient:

$$r_{xy} = \frac{\sum\limits_{i=1}^{n}(x_i - \bar{x})(y_i - \bar{y})}{(n-1)s_x s_y} \tag{8.8}$$

where $\bar{x}$ and $\bar{y}$ are the means of the two samples and s_x and s_y their standard deviations. If the *x* values tend to be smaller than $\bar{x}$ when the corresponding *y* values are greater than their mean, and vice versa, the numerator becomes negative. This means that, if *y* tends to decrease when *x* increases, the correlation coefficient becomes negative. If *x* and *y* tend to vary in the same direction the coefficient becomes positive. Dividing by the standard deviations ensures that the value of the coefficient always lies between -1 and $+1$. This means that a correlation coefficient that is close to $+1$ indicates a strong linear association, or correlation, between the samples, whereas a coefficient close to zero indicates a weak correlation or no correlation at all. If the coefficient is close to -1 we say that there is a negative correlation between the samples.

Figure 8.11 shows example scatterplots of *x*- and *y*-samples that all have some sort of relationship to each other. Each plot contains a correlation coefficient. The first plot gives

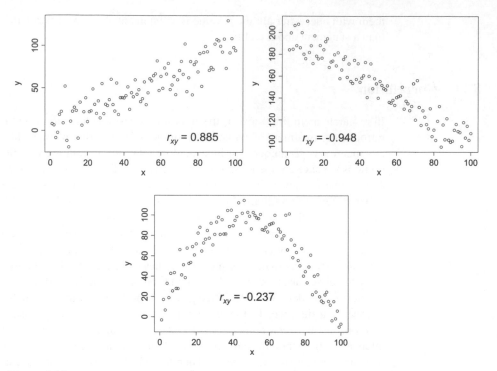

Figure 8.11
Examples of data sets where there is some type of relation between the *x*- and *y*-variables.

an example of strong positive correlation. In this diagram, the points seem to fall along a straight line with positive slope, though there is significant scatter about the line. In the second plot the slope is negative and, consequently, the correlation coefficient indicates a strong negative correlation. Although there is a clear relationship between the variables in the last plot, the coefficient tells us that there is little or no correlation between them. This demonstrates the important fact that a lack of correlation is not the same thing as lack of relationship – only that there is no *linear* relationship between the variables. For this reason, it is important to always plot the data to avoid the risk that potential relationships among variables remain undiscovered.

It is also important to point out that, although we use correlation analysis to detect prospective relationships among variables, a high degree of correlation does not prove that there is a direct relationship. If we, for instance, were to compare sales data for sunglasses with the number of people who show symptoms of hay fever, we would probably obtain a positive correlation. The same would be true for a comparison between the occurrence of ski accidents and mitten sales. This does not mean that sunglasses are causing hay fever, or that wearing mittens makes skiing more dangerous. Such apparent relationships may occur by coincidence or, as in this case, if the two variables have a common dependence on a third variable.

The point is that statistical analyses are not a substitute for subject matter knowledge. We should rather use our knowledge of our research fields as much as possible, as this makes it much easier to understand what the data are telling us. Finding a strong correlation is of little use if we cannot explain why the variables should be related to each other.

8.12 Regression Modeling

Regression is a method for building mathematical models of how a response variable is affected by changes in one or several predictor variables. The first step in regression is to choose the form of model that we want to use. As a general rule, we always strive to use the simplest form that is appropriate. This rule is of general importance in research and science and, as explained in Chapter 5, it is called the principle of parsimony. Scientists should always prefer the simplest possible explanation of any given phenomenon and not add any unnecessary assumptions or details. Complex explanations often indicate that there is a problem with the basic idea. The tenor of the principle is well captured in a phrase that is often credited to Albert Einstein: "Simplify as much as possible – but not more!" We will return to this discussion at the very end of this section.

The simplest model that we can use is the straight line, so let us start with that. Figure 8.12 shows a scatter plot where each black diamond corresponds to one value on the x-axis and one on the y-axis. We say that the y sample is plotted versus the x sample. It seems like y has a reasonably linear relationship to x, so fitting a straight line through the data is, in this case, appropriate. We are all familiar with the equation of the straight line:

$$y = a + bx \tag{8.9}$$

It contains two parameters that we need to estimate from the data: a is the intercept with the y-axis and b is the slope of the line. In regression, these parameters are estimated using the method of least squares. Complicated as this may sound it is, in principle, very simple. All we have to know is that the fitted line always goes through the coordinate $(\bar{x}, \bar{y})$, which is marked with a hollow circle in Figure 8.12. The parameters are determined by rotating the line about this point until it is as close to the points as possible.

Mathematically, this corresponds to minimizing the residuals. Residuals were introduced in Chapter 7 as the distance between the observations and the sample mean. This definition is appropriate when the variable is only expected to display random variation about its

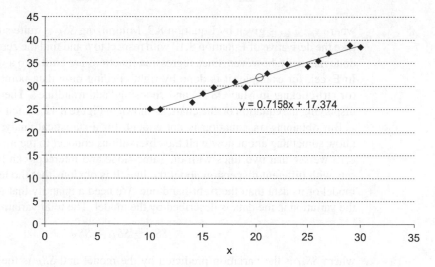

Figure 8.12
A scatter plot with a linear regression line and its regression equation, as obtained in Excel.

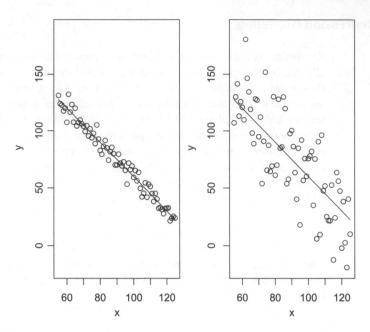

Figure 8.13
Two lines that have the same slope and intercept, despite being estimated from data sets with different levels of noise. It is obvious that the left-hand line is a better model of its data than the right-hand one. (Adopted from Crawley [4].)

mean. Here, we expect y to change when x changes so, in regression, the residuals are the distances between the observations and the *model*. The sum of squared residuals is, therefore:

$$SS_R = \sum \left(y_{observed} - y_{predicted}\right)^2 = \sum_i (y_i - a - bx_i)^2 \qquad (8.10)$$

where $y_{predicted}$ is given by Equation 8.9. Minimizing SS_R is quite straightforward. We just take the derivative of Equation 8.10 with respect to b and find the zero-point. This is seldom done manually, however, since regression functions are built into a wide range of software. In Excel, for example, it is done by right-clicking on a data point in an x–y scatter plot (or ctrl-clicking in Mac OS X) and choosing "add trendline". There is also an option for displaying the equation on the chart, as shown in Figure 8.12, which was produced in Excel.

But obtaining an equation is not enough for a statistical analysis. It is also useful to know something about how well the observations connect to the model. Looking at Figure 8.13 we see that two lines with the same slope and intercept can be estimated from data sets with different dispersion about the line. It is obvious that the left-hand line is a better model of its data than the right-hand one. We need a quantity that measures how much of the variation in the data is described by the model. The total variation in the data is:

$$SS_T = SS_M + SS_R \qquad (8.11)$$

where SS_M is the variation predicted by the model and SS_R is the contribution from the residuals. SS_R is calculated from the distances between the observations and the predictions, whereas SS_M is based on the distances from the predictions to the grand mean. We will

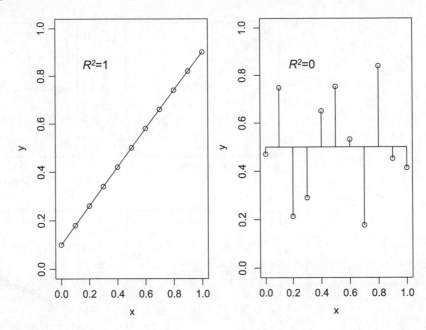

Figure 8.14
Illustration of a perfect fit ($R^2 = 1$) and complete lack of fit ($R^2 = 0$). (Adopted from Crawley [4].)

introduce a parameter called the coefficient of determination, R^2, which is the portion of the total variation that is described by the model:

$$R^2 = \frac{SS_M}{SS_T} = 1 - \frac{SS_R}{SS_T} \tag{8.12}$$

As statistical software usually provides R^2 along with the regression coefficients, it is useful to know how to interpret it. If the line connects perfectly to all the observations we have a perfect fit and $SS_R = 0$. In this case, $R^2 = 1$. If there is only random variation in the data, the model explains none of the variation and $SS_M = 0$. In this case, $R^2 = 0$. These two extreme cases are displayed in Figure 8.14. Needless to say, it is desirable to have an R^2 close to one, since this means that the model is an accurate representation of the data. Be aware, however, that you can always increase R^2 by increasing the complexity of the model. By adding quadratic and higher order terms to Equation 8.9 you may, theoretically, be able to make the model follow the observations perfectly. This is *not* desirable, since there is always some noise in the data. We will not obtain a useful model if we fit it to the random variation.

There are also other aspects of the fit that are important to assess. If a model describes the variation in the data well, the observations should be scattered randomly around it. More specifically, the residuals should be normally distributed and have constant variance. It is useful to make a few residual plots to make sure that this is the case, as shown in Figure 8.15. The first one shows a data set and a linear regression model fit to it. The next two show the residuals versus the observation number and versus the fitted value. If the model is good, these plots should display no structure at all. If, for example, the variance tends to decrease with observation number this could be a sign that the measurements were made during a run-in phase, before the process under study had reached stability. Residuals that

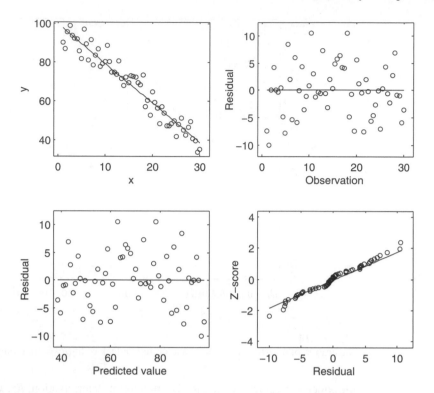

Figure 8.15
A linear model and three residual plots: residuals versus observation number, versus fitted value, and, finally, a normal plot of the residuals.

show a trend as function of the fitted value is a sign that the wrong model has been chosen. There could, for instance, be a curvature in the data that is not captured by the model. If so, the residuals will tend to be positive in some regions of the plot and negative in others. The last plot is a normal plot of the residuals. If they are reasonably normally distributed, they plot on a straight line, as in this diagram.

There are of course more complex relationships than the straight line. There could, for instance, be more than one predictor variable. A linear regression model with two predictors has the general form:

$$y = \beta_0 + \beta_1 x_1 + \beta_2 x_2 + \beta_3 x_1 x_2 + \beta_4 x_1^2 + \beta_5 x_2^2 + \varepsilon, \tag{8.13}$$

where the x values are the predictors and the β values are called regressors. The last term, ε, represents the regression error. When there are two or more predictors we talk about a *multiple* linear regression model. It is worth noting that it is called a linear model even if it contains quadratic terms, because the response variable is modeled as a linear combination of the regressors. Statistical software basically fits multiple linear regression models to data in the same way as the straight line was fit.

We use the same criteria to evaluate multiple linear regression models as above: a high R^2 is desirable and the residuals should be normally distributed and have constant variance. In some cases it may be difficult to obtain a model that meets these criteria. Sometimes this is a sign that the response variable has a more complex dependence on one or several of the predictor variables than the linear model can describe. In many cases, such problems can be remedied by variable transformations. If we expect, for instance, that our y has a logarithmic relationship to x, it may be modeled using a linear model where x is replaced by the logarithm of x. In severe cases we may have to resort to non-linear regression models, which are often supported by statistical software.

When working with statistical models we should keep in mind that the model is a representation of reality. As such it should both provide a good illustration and a good prediction of reality. As already mentioned, there is often a trade-off between these two aspects of the model, as a complex model may minimize the residuals without giving a good illustration of the important relationships in the data. For this reason it is important to keep in mind that statistical software often provides us with a maximum model – a complex model that contains the largest possible number of terms. The terms are often presented with associated p-values. We will get a more parsimonious model if we remove the terms with the largest p-values one by one, while not letting R^2 decrease too much. The reduced model leaves us with terms that actually explain the observed relationships in the data, and thereby provide a good illustration of reality. When reducing the model, we should begin with higher order terms and cross terms, and keep an eye on R^2 as we proceed. Parsimony requires the model to be as simple as possible, while retaining enough accuracy to be useful.

These and other ideas from this chapter will be applied in the next chapter, which deals with experimental design. We will also see how multiple linear regression models may be fit to data in Excel.

8.13 Summary

- A statistical hypothesis test answers a question that contains a statistical parameter and a word expressing a difference. "Is the sample mean different from the target value?' is an example of such a question.
- The null hypothesis assumes that any apparent difference is due to sampling error – natural, random variation in the data. It is coupled to a reference distribution that describes all possible outcomes of the experiment if it is completely governed by chance.
- To ensure that this reference distribution is valid, the order of the experimental treatments must be randomized. Otherwise, drift in the conditions could produce apparent effects that would be wrongly attributed to the experimental treatments.
- The alternative hypothesis is the opposite of the null hypothesis and generally represents an effect that we want the experiment to reveal. If the significance test discredits the null hypothesis, the alternative hypothesis is supported and we say that the result is *statistically significant*.
- The confidence level is the risk of making a Type I error, or the probability of rejecting the null hypothesis when it is true. The power is the risk of making a Type II error: accepting the null hypothesis when it is false.
- The one-sample t-test is used for comparing a sample mean to a target value. The two-sample t-test is used for comparing the means of two independent samples to each other. If the samples are dependent, the paired t-test is used, which is a special case of the one-sample t-test.
- Three or more sample means are compared using ANOVA. It is based on breaking the variation in the data down into several components and comparing them to each other. If the variation in the data is due to two variables (or *factors*) the two-way ANOVA is used.
- The correlation coefficient is a measure of the strength and direction of the linear association between two variables. The existence of a strong correlation between two variables

may indicate a direct dependence between them, but it is important to keep in mind that correlation is not the same thing as causation.

- A regression model can be used to describe the relationship between numerical variables. It is a mathematical function fitted to the data. The coefficient of determination, R^2, is a measure of how well the model describes the variation in the data. Residual plots are used to assess the quality of the model. The residuals should be normally distributed and have constant variance.

Further Reading

Statistics for Experimenters by Box, Hunter and Hunter [3] introduces a large number of useful hypothesis tests using practical examples and is highly recommended as a reference book for experimenters. It also treats statistical modeling. Two other books that provide good introductions to statistical modeling are *Statistics: an introduction using R* by Crawley [4] and *Multivariate data analysis* by Hair, Anderson, Tatham, and Black [5].

Answers for Exercises

8.1 A mean fuel consumption of 5.84 liters per 100 km is needed to provide a significant result at the 95% confidence level.

8.2 Reading from the table of probability points for the t-distribution in the Appendix, $t_{0.05,4} = 2.132$. The one-sided confidence interval lies below 6.14.

8.3 $t_{obs} = 2.064$. The t-table in the Appendix gives $t_{crit} = t_{0.05,5} = -2.015$. As $t_{obs} \gg t_{crit}$ the null hypothesis is strongly supported. (Note that t_{crit} is negative due to the "$\geq$" in the null hypothesis, which means that it will be discredited by observations in the left tail of the distribution, and that forgetting the minus sign would reverse the conclusion.)

8.4 The mean difference is 0.08. Using the pooled standard deviation this yields a t_{obs} of 0.8. The critical t-values for this two sided test are $t_{0.025,8} = \pm 2.306$. Since t_{obs} is within these boundaries H_0 is supported: the data suggest that there is no significant difference between the summer and winter fuel consumption.

8.5 The ANOVA table should look like the following. There is a difference between the treatments that is significant at the 95% confidence level.

SOURCE	v	SS	MS	F	p
Treatment	2	4.906	2.453	5.56	0.027
Error	9	3.969	0.441		
Total	11	8.875			

8.6 The ANOVA table should look like:

SOURCE	v	SS	MS	F	p
Operator	1	0.0080	2.0080	6.720	0.017
ECU	1	0.0040	0.0040	3.360	0.081
Error	21	0.0250	0.0012		
Total	23	0.0360			

References

1. Fisher, R.A. (1951) *The Design of Experiments*, 6th edn, Oliver and Boyd, Edinburgh.
2. Huitson, A. (1966) *The Analysis of Variance*, Charles Griffin & Co., London.
3. Box, G.E.P., Hunter, J.S., and Hunter, W.G. (2005) *Statistics for Experimenters: Design, Innovation, and Discovery*, 2nd edn, John Wiley & Sons, Inc., Hoboken (NJ).
4. Crawley, M.J. (2005) *Statistics: An Introduction Using R*, John Wiley & Sons Ltd, Chichester.
5. Hair, J.F., Anderson, R.E., Tatham, R.L., and Black, W.C. (1998) *Multivariate Data Analysis*, 5th edn, Prentice-Hall, Upper Saddle River (NJ).

9 Experimental Design

Sometimes the only thing you can do with a poorly designed experiment is to try to find out what it died of.

—Ronald A. Fisher

In the last chapter we saw how to use statistical methods to find correlations and differences in data. Chapter 6 showed that experimental method helps us to connect effect with cause. Combining experimental and statistical method makes it easier to answer scientific questions both more economically and with better precision than otherwise. In this chapter, statistical approaches will be put into an experimental context. After a brief look at single factor experiments we will look at the very important scenario where several variables are suspected to simultaneously affect a response. The purpose is to explain some basic thoughts and introduce some classic designs using practical examples. Exercises should be completed when encountered in the text to confirm that you understand the methods before proceeding. Further reading is suggested at the end for those who want a more comprehensive treatment.

9.1 Statistics and the Scientific Method

We saw in Chapter 6 that many experiments do not involve statistical tests at all. The interesting effect may be so strong that it is not necessary to use statistics to separate it from the noise. In many cases, however, statistical tools are indispensable components of the experimental strategy, especially when effects are weak or when background factors are important. The combination of statistical and experimental method is called *design of experiments*. Fisher [1] originally used the term for planning an experiment with a particular significance test in mind. The teatime experiment in the last chapter is a good example of this, borrowed from his book *The Design of Experiments*. It shows how an experiment can be designed to discern the potential weak effect of a single factor by applying replication and randomization. Today, as we shall see, the term "design of experiments" has been extended to include orthogonal multifactor experiments where the analysis is often based on linear models of the data rather than pure significance tests.

When significance tests are used, it is important not to confuse the statistical hypothesis with the scientific one. Scientific hypotheses go beyond particular sets of data and make general statements about the world. As explained in Chapter 6, they may even be a basis for explanatory theory if they involve a mechanism. Statistical hypotheses are not explanatory; they only make simple statements about mathematical relationships. They may suggest that the mean of one sample is greater than that of another but make no statements about the

Experiment!: Planning, Implementing and Interpreting, First Edition. Öivind Andersson.
© 2012 John Wiley & Sons, Ltd. Published 2012 by John Wiley & Sons, Ltd.

reason behind this difference. While statistical hypotheses can be important in providing support for a more complex scientific hypothesis, they say very little in themselves about the workings of the world. We should also remember that individual tests rarely provide sufficient information to support a final conclusion about the truth of a scientific hypothesis. To have generality, results should be consistent under a wide variety of circumstances. The lack of statistical significance in one experiment is not proof of the lack of an effect.

In summary, statistical tests are tools that may help researchers draw conclusions that would otherwise be difficult to support but they rarely answer research questions directly. The statistical question has to do with the potential difference between specific samples, whereas the research question concerns general patterns and regularities. Statistics is a study of numbers and science is the study of the world. The reason that they often occur together is that the world can often be studied with the help of numbers.

9.2   Designs with One Categorical Factor

Experimentation is to create a specific condition and investigate its effect on a specific outcome. We may be interested in how a surface treatment affects the properties of a material, how a solvent affects the yield of a synthesis or how a drug treatment affects a medical condition. The input variables are then called categorical factors, since they describe states that can be labeled but not ordered with respect to each other: we either use one treatment or another, or possibly none at all. In such cases, the hypothesis tests introduced in the last chapter can be used to analyze the results.

If the natural variation in the measured variable is large and the samples are small, it may be difficult to discern the effect of the treatment from the noise in the data. In such cases it is often necessary to design the experiment with noise and background factors in mind. This need is common in medical studies, for example, since patients are affected by a wealth of social, environmental and other background factors that are not under the experimenter's control.

The most common method for avoiding background effects is to make controlled experiments. This means that two groups are compared, where one is exposed to the experimental treatment and the other does not receive any treatment at all. If the experimental group shows an effect and the control group fails to do so, this indicates that the effect is due to the treatment. A background factor that may still be important in such studies is that the mere *expectation* of an improvement may produce a response in the experimental group. If the control group is aware that it does not receive treatment this effect will be absent among them. For this reason they should be given a placebo, which is a simulated treatment that has no effect in itself. The study should be *blinded*, meaning that the control group is to be kept unaware of the fact that they receive a placebo. In some cases it may even be expected that the experimenters' expectation of an effect will cause them to act differently with the two groups and this, too, could induce a false response. It is then better if both the experimenters and the subjects are kept unaware of which group receives the placebo, and this is called a *double-blind* study. To minimize the risk that other background factors bias the outcome, the subjects should be randomly allocated to the different treatment groups. Such experiments are collectively referred to as *randomized controlled experiments*. The data are analyzed using a two-sample *t*-test or, if the procedure is generalized to more than one treatment, using ANOVA. It is important to use sufficient sample sizes to decrease the risk of Type II errors. As discussed in the last chapter, an appropriate sample size is found by power analysis.

When discussing the *t*-test in the last chapter we differed between dependent and independent samples. With independent samples we simply compared the sample means to each other. In dependent samples, however, one individual is associated with one observation in each sample. By investigating the *difference* between these paired observations we remove the variation between individuals. This increases the precision in the data compared to independent samples. The analysis is made using the *paired t*-test.

Randomized controlled experiments are often considered to be one of the most rigorous test procedures in medical and similar studies but, since they are based on independent samples, the precision can be improved. Variation between the individuals may obscure the effect of a treatment. If we instead design the experiment to study the difference induced by the treatment in single individuals, we may analyze the results using the paired *t*-test. If you need to review why this procedure increases the precision in the data, you could have another look at the discussion about the dietary data in Table 8.5.

This procedure is used in an experimental design called a *crossover experiment*. The subjects are then exposed to a sequence of at least two different treatments, one of which may be a placebo. As described in Figure 9.1, the study has two arms and the subjects should be randomly allocated to each arm. The design is most often balanced so that all subjects are exposed to the same number of treatments and treatment periods. We may, for example, be interested in finding out if a certain substance decreases the LDL cholesterol level in the blood. Before treatment we measure the baseline cholesterol level in each individual. The subjects in one arm of the study then receive the substance for a period of, say, 15 days. At this point the cholesterol level is measured again. The next step is to cross this group over to the other treatment, which could be a placebo. A "washout" period is usually employed to avoid carry-over effects that may arise if the response remains in the subjects for some time after the treatment. After receiving the placebo for 15 days the cholesterol level is measured anew. Subjects in the other arm of the study are treated in exactly the same way but they receive the placebo first and the substance of interest last.

So, compared to a randomized controlled experiment, what have we gained by this more complex procedure? The most obvious advantage is that each individual is his or her own control. Comparing the cholesterol level before and after treatment in each person we remove the baseline variation between the individuals, exactly analogous to the dietary experiment in Chapter 8. This is appropriate because we are of course interested in the change induced by the substance and not in how the level varies between individuals before the treatment. If we were to exchange the placebo for a second type of treatment, this design also allows us to analyze the potential effect of the *order* of the treatments.

In crossover designs, the variation from the baseline can be analyzed using a paired *t*-test. In our case this test both gives the effect of the treatment and the placebo effect. If we are interested in the effect of the order in which two treatments are given, a two-way ANOVA

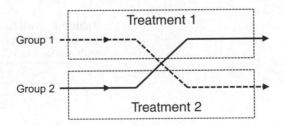

Figure 9.1
Schematic description of a crossover experiment.

can be used with the order and the treatment as the two factors. Again, randomization is often necessary to avoid bias in the data; a power analysis should be made to determine an appropriate sample size.

9.3 Several Categorical Factors: the Full Factorial Design

We frequently expect the response in an experiment to be affected by several factors. If so, it is interesting to find out which factor has the strongest effect on the response. Another important question is if the effects are separate or if the way in which the factors are combined is important. These questions can be answered by full factorial experiments.

We will begin by a case with only two factors, since this is convenient for understanding. The design is shown in Table 9.1, where the columns represent the two factors and the rows correspond to the investigated factor combinations, or treatments. We may note that all the possible combinations are present. To understand how the effects of the factors still can be separated, we will introduce the concept of *orthogonality*. This requires a brief mathematical digression, but we will soon be back on track again.

Mathematicians discern between two types of quantities: scalars and vectors. Numbers in general only have size and are called scalars. Scalars are used to describe things like the outside temperature or the atmospheric pressure. Vectors have both size and direction and are expressed in terms of coordinates. The wind velocity is one example. The plane in Figure 9.2 is defined by two vectors, x and y, each with a length of one unit. They represent the two principal directions in which it is possible to move in that plane. Any vector in the plane can be expressed in terms of (x,y) coordinates. The coordinates of the x-vector are $(1,0)$, the y-vector is expressed as $(0,1)$ and the vector a has the coordinates $(1,1)$. All these coordinates can be seen as instructions for getting from the origin to a certain point in the x–y plane: x says "move one unit in the x-direction", y says "move one unit in the y-direction" and a says "move one unit along x and then one along y". The last instruction is just the combination of the first two and, for this reason, a is said to be a *linear combination* of x and y. Specifically, we may obtain a by adding the x- and y-coordinates of x and y separately, since $(1 + 0, 0 + 1)$ gives us $(1,1)$. As x and y are at right angles to each other they are, by definition, orthogonal. Since a can be obtained by combining the other two, it is not orthogonal to any of them. In mathematical terms, a set of orthogonal vectors is characterized by the fact that none can be obtained by a linear combination of the others.

If two vectors are orthogonal the inner product between them is zero. We obtain the inner product by multiplying the respective elements of the two vectors and summing up the products. For the orthogonal x- and y-vectors this gives $(1 \times 0) + (0 \times 1) = 0$. If we

Table 9.1 Two-level full factorial design with two factors.

A	B
−1	−1
1	−1
−1	1
1	1

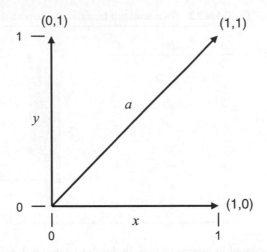

Figure 9.2
Plane defined by two vectors, x and y. The third vector a is a linear combination of the two.

apply this calculation to the two columns of Table 9.1 we see that they too are orthogonal to each other:

$$(-1 \times -1) + (1 \times -1) + (-1 \times 1) + (1 \times 1) = 0 \qquad (9.1)$$

Now, what on earth does this have to do with experimental design? Well, in a multifactor experiment we want to separate the effects of the factors on a response variable and we may do this by utilizing an orthogonal test matrix. As we have seen, the design in Table 9.1 is one such example, since the combination of settings in each column is orthogonal to the other. The easiest way to understand how this works and why it is useful is to show a simple example.

Imagine that you are interested in three categorical factors, A, B and C. They could, for instance, be factors of interest for the amount of wear on a bearing surface. A could be a manipulation of the surface of the shaft, B a manipulation of the bearing, and C an additive in the lubricant. Let us say that you want to investigate the amount of wear that you get with and without these factors. This means that each factor will have two settings in the test matrix: "with" and "without". We will simply call these settings $+1$ and -1. Table 9.2 shows the test matrix for all possible combinations; the levels are denoted by "+" and "−" signs for simplicity. The design is also shown graphically as a cube.

The vertical columns correspond to different factors and the horizontal rows to different treatments. In this case the treatments are combined so as to obtain orthogonal columns. We may confirm this by verifying that the inner products between the columns are zero.

> **Exercise 9.1:** Confirm the orthogonality by entering the table into a Microsoft Excel® spreadsheet and calculating the inner products of the columns. This is most conveniently done using the Excel function SUMPRODUCT. (Note that you must write "1" and "−1" instead of "+" and "−" in the spreadsheet.)

The orthogonal test matrix is obtained by filling out the first column with alternate "+" and "−" signs. The second column is obtained in a similar way but here the signs are

Table 9.2   Two-level full factorial design with three factors.

A	B	C
−	−	−
+	−	−
−	+	−
+	+	−
−	−	+
+	−	+
−	+	+
+	+	+

repeated in groups of two. In the third column the signs appear in groups of four. We may expand this test matrix to any number, n, of factors, with the groups of signs doubling in size for each consecutive column. The three factors in Table 9.2 thereby give us $2^3 = 8$ rows and, for this reason, this is sometimes called a 2^3 design. The ordering of the treatments is called *Yates order* and the resulting test matrix is called a *full factorial design*.

In addition to providing an orthogonal test matrix the full factorial design tests all possible combinations of factor levels. As the design in Table 9.2 tests each factor at a high and low setting it is called a *two-level* full factorial design. The design can be extended to more levels. For a three-level design we usually denote the levels by "−", "0" and "+" signs. With three factors the number of rows in the matrix becomes $3^3 = 27$. In general, the number of treatments, k, in a full factorial design is given by:

$$k = l^n \tag{9.2}$$

where l is the number of levels and n is the number of factors. Figure 9.3 shows a graphical comparison of a two-level and a three-level full factorial design with three factors.

Let us look at some data to see why the orthogonality is convenient when analyzing a full factorial experiment. In Table 9.3 we have taken the test matrix in Table 9.2 and added factor interactions by multiplying the three columns A, B and C with each other. The resulting

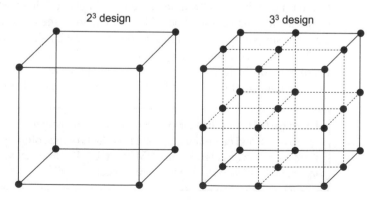

Figure 9.3
Comparison of a two-level and a three-level full factorial experiment.

Table 9.3   The full factorial design in Table 9.2 supplemented with columns for interactions.

	A	B	C	AB	AC	BC	ABC	RESPONSE
	−1	−1	−1	1	1	1	−1	56.2
	1	−1	−1	−1	−1	1	1	55.6
	−1	1	−1	−1	1	−1	1	61.6
	1	1	−1	1	−1	−1	−1	52.2
	−1	−1	1	1	−1	−1	1	54.0
	1	−1	1	−1	1	−1	−1	50.0
	−1	1	1	−1	−1	1	−1	60.3
	1	1	1	1	1	1	1	51.1
Divisor:	4	4	4	4	4	4	4	
Effect:	−5.8	2.35	−2.55	−3.5	−0.8	1.35	0.9	

matrix contains columns corresponding to the three factors, three two-way interactions and one three-way interaction. Note that these interactions do not have to be considered during the experiment; they are calculated afterwards. The column at the far right is a measured response that results from the settings of A, B and C. It could, for instance, be the outcome of an abrasive wear test. If you wonder about the numbers below the table their meaning will soon become clearer.

> **Exercise 9.2:** Enter the data in Table 9.2 into an Excel spreadsheet. Then add the interaction columns by multiplying the columns with each other. This is done by multiplying the top elements and double clicking the fill handle of the output cell. You should obtain interaction columns identical to those in Table 9.3. Finally, enter the response data in the last column.

Now, if we take the inner product between a factor column and the response column we get a measure of the linear association between them. If there is no relationship between the factor and the response they are linearly independent. In other words, they are orthogonal to each other and the inner product will be zero. Similarly, a large inner product translates into a strong association between the factor and the response. This is because taking the inner product between two vectors is very similar to calculating their correlation coefficient, as described in Equation 8.8. Since the factor columns are linearly independent of each other this amounts to a method for complete separation of each factor's effect on the response. It goes without saying that this is very useful to an experimenter: we design an experiment where we vary all factors simultaneously but, still, their effects on the response can be completely isolated.

Recall that if the test plan were not orthogonal, one column could be obtained from a linear combination of the others. This means that the factors' effects on the response would be mixed with each other. This is why it is difficult to see clear relationships in observational studies where the data never originate from orthogonal test plans. In observational data, a multitude of factors vary haphazardly and simultaneously and this causes their effects to bleed into each other. This problem is called *collinearity*, a word describing that the factor columns are not linearly independent.

Continuing with the numbers below Table 9.3, we will first calculate what is called the *effects* of the factors. The effect measures the magnitude of the response that results from a change in a factor. Looking at the first column we see that there are four measurements at the low setting of A (-1) and four at the high setting $(+1)$. One way to determine the effect of factor A is to take the difference between the average responses at the high and low settings. In other words, it is the difference between the average responses on the two shaded surfaces of the cube in Table 9.2. This difference will say something about whether the response tends to increase or decrease when A increases from -1 to $+1$ and, if so, how much. The effect of A is explicitly calculated as:

$$\frac{55.6 + 52.2 + 50.0 + 51.1}{4} + \frac{56.2 + 61.6 + 54.0 + 60.3}{4} = -5.8 \qquad (9.3)$$

We recognize the denominator as the divisor in Table 9.3, as well as the effect value given there. Factor A, the surface manipulation, seems to be effective since the amount of wear decreases when it is applied. Rearranging the terms we see that this calculation is exactly equivalent to taking the inner product between the A and response columns and dividing by four. The reason that the divisor in Table 9.3 is four is simply that this is the number of measurements at each factor level. In summary, the effect of A measures the average change in the response when A changes from the low level to the high level. Since this calculation only detects the response to a single factor it is called a *main effect*.

The two-way interactions show if the response to a change in one variable is affected by the setting of another. Let us look at the BC interaction in Table 9.3. With the data in this column we can first calculate the effect of B when C is at the high level; that is, using the four responses that were measured when C was set to $+1$. Similarly, we can calculate the effect of B when C is set to the low level. Now, the difference between these two effects tells us if the setting of C affects the response to a change in B. The interaction is explicitly calculated as:

$$\frac{1}{2}\left\{ \left(\frac{60.3 + 51.1}{2} - \frac{54.0 + 50.0}{2} \right) - \left(\frac{61.6 + 52.2}{2} - \frac{56.2 + 55.6}{2} \right) \right\} = 1.35 \qquad (9.4)$$

Expressed in words, this is half the difference between the effect of B when C is at the high level and the effect of B when C is at the low level. Take a minute to think about this. If the interaction effect is zero this means that the response to a change in B is the same regardless of the setting of C – there is no interaction between them. If the interaction effect is positive it means that the response to a change in B is stronger when C is at the high setting. A negative interaction effect means that the response to a change in B is stronger when C is at the low level. In extreme cases, B may even have the opposite effect on the response at the low setting of C. All of these types of interactions frequently occur in real data. Again, rearranging the terms in Equation 9.4 we see that it is exactly equivalent to taking the inner product between the BC and response columns and dividing by four. As a rough rule of thumb, main effects tend to be largest, two-way interactions tend to be smaller, while three-way and higher interactions usually can be ignored. Looking at the effects calculated at the bottom of Table 9.3 this rule is roughly confirmed, although the AB interaction is relatively large. The rule becomes more manifest in a larger experiment.

Exercise 9.3: Use your Excel spreadsheet to calculate the main and interaction effects in the data in Table 9.3. Take the inner product between each column and the response column and divide by four. This should result in the effects given in Table 9.3.

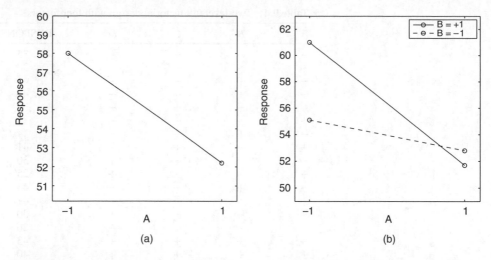

Figure 9.4
(a) Effect plot of factor *A* in Table 9.3. (b) Interaction plot of the *AB* interaction.

It is always convenient to present results graphically, since it is easier to see relationships in diagrams than in numbers. Data from full factorial experiments are often displayed in effect plots and interaction plots. Figure 9.4*a* shows an effect plot for factor *A* in Table 9.3. It simply plots the average responses at the low and high levels of *A* with a line between the points. Figure 9.4*b* is an interaction plot of the *AB* interaction in Table 9.3 and may require a little more explanation. It could be said to contain two effect plots for factor *A*. One effect is measured at the high setting of *B*, the other at the low setting, as indicated in the legend. Since the two lines have different slopes, a single glance shows us that there is an interaction effect and that it is fairly strong. The setting of *B* clearly affects the response to *A*.

We have now learned how to calculate the effects in a full factorial experiment but an important part of the analysis is still missing. After reading the last two chapters we know that measurements always suffer from noise and we need methods to separate the effects from this background variation. Some effects will be strong compared to the noise and some will be weaker, but all effect calculations will return numbers. The problem is to understand which numbers show the random variation and which measure real effects.

We can often find this out graphically by using the now familiar normal plot (Chapter 7). Since the effects are calculated from orthogonal vectors they are independent and can be treated as a random sample. Sorting the effects from the smallest to the largest and plotting them versus their normal scores will produce a normal plot. If there were no effects in the data, the calculated effects would only measure the experimental error. If so, they would be normally distributed and plot on a straight line in the diagram. We hope, of course, that some effects are large compared to the error. If so, they will plot off the line. A relatively large sample is needed to produce a neat normal plot and we will, therefore, move to a two-level four-factor experiment for illustration. The factors *A*, *B*, *C*, and *D* in Table 9.4 could be any categorical variables suspected of affecting a continuous response variable. They are given in Yates order to make it easier for you to enter them into an Excel spreadsheet. To calculate the interaction effects, the table must be extended to include the six possible two-way interactions (*AB*, *AC*, *AD*, *BC*, *BD*, and *CD*), the four possible three-way interactions

Table 9.4 Two-level full factorial design with four factors.

A	B	C	D	RESPONSE
−	−	−	−	35
+	−	−	−	30
−	+	−	−	45
+	+	−	−	41
−	−	+	−	35
+	−	+	−	31
−	+	+	−	44
+	+	+	−	41
−	−	−	+	30
+	−	−	+	25
−	+	−	+	44
+	+	−	+	41
−	−	+	+	30
+	−	+	+	26
−	+	+	+	43
+	+	+	+	40

(ABC, ABD, ACD, and BCD), and the four-way interaction. In total, this gives us 15 main and interaction effects, which should be sufficient to produce a normal plot.

> **Exercise 9.4:** Enter the data in Table 9.4 into an Excel spreadsheet and extend it to include all the interaction columns. Calculate the main and interaction effects. The divisor should simply be the number of measurements at each setting. To obtain the normal plot, transfer the calculated effects to a separate sheet, order them from smallest to largest (using the "Sort A-Z" function) and calculate the standard normal scores according to the procedure in Example 7.4. If you succeed, the normal plot should look like that in Figure 9.5.

Calculating the effects we see that some are larger than others but we do not easily see which effects stand out from the noise. When plotting them as in Figure 9.5 it only takes a single glance to see that four effects are greater than the noise on the straight line. The largest effect is due to factor B, which plots at the far right of the graph. This means that the response increases when B increases. The second largest effect is due to factor A on the left. This effect is negative, meaning that the response decreases when A increases. To put it simply, factors plotting to the right are positively correlated with the response, while those plotting to the left are negatively correlated. The other two significant terms are D and the BD interaction.

Strictly, it is not correct to talk about correlation when the factors are categorical, so keep in mind that we refer to the *coded* values −1 and +1, which were arbitrarily assigned to the factor levels. Another way of separating the statistically significant effects from the noise is to apply a *t*-test to the effects. This method will be introduced later in this chapter, in connection with continuous factors and regression.

Before finishing this section we should discuss some experimental precautions that are commonly used in full factorial experiments. First of all, a full factorial experiment should never be run in Yates order. Factor D in Table 9.4, for example, would then be at the low setting during the entire first half of the experiment and at the high setting during the second

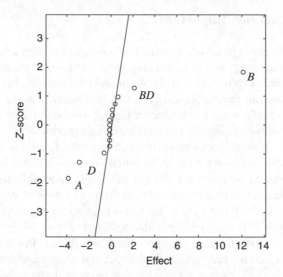

Figure 9.5
Normal plot of effects calculated from Table 9.4.

half. There could be a background factor that drifts slowly during the measurements, for instance if they are made during a run-in phase. If the response were to increase due to this drift the average response would be lower during the first measurements. This, in turn, means that factor D would show a false effect due to this drift. To safeguard against false background effects, the run order should always be randomized. It is quite easy to produce a column of random numbers alongside the test matrix in Excel, for example using the RAND or RANDBETWEEN functions. The test matrix can then be randomized by sorting all the columns according to these random numbers.

We also need to estimate the experimental error. This is normally not done using the graphical approach in Exercise 9.4 but by using regression analysis. Recall from Chapter 8 that it costs one degree of freedom to estimate a parameter in a regression model and that additional degrees of freedom are needed to estimate the error. This means that the table with factor and interaction columns must be taller than it is wide, since each column is associated with one parameter. Such a table is obtained by replicating the measurements one or several times. To avoid background effects, the replicates should be run in random order and be mixed with the original measurements. You may note that none of the examples in this section contain replicates. This is to facilitate the explanation of the principles of the full factorial design but, in real life, your full factorial experiments should never look like this. They should always include at least one replicate of each measurement.

Another common precaution in this type of experiment is blocking. Sometimes we have identified a background factor that could, potentially, affect the response. In a chemical synthesis, for example, we may have a limited supply of a certain raw material. If we need to take this material from two separate batches, this could affect the result. We may find out if it does by simply putting the batch into the design as a separate factor. The batch is then said to be a *block variable*. If it does not affect the response, the runs with one batch are equivalent to those with the other. They can then be used as replicates of the other runs and this provides us with the extra degrees of freedom needed to estimate the error.

9.4   Are Interactions Important?

Many of us have learned that, when experimenting with several factors, we must investigate one factor at a time, keeping all others fixed. This method makes it all but impossible to detect factor interactions. A common response to this is that interactions are not important and, in the rare cases when they do matter, we can get rid of them by suitable variable transformations. Statistician George Box inimitably dismisses of these statements in the following, imaginary rabbit breeding experiment [2].

If we do not know anything about breeding rabbits we may start by an appropriate control case with no rabbits in the hutch, to confirm that this case produces no rabbits. Adding a doe to the hutch we find that this single factor fails to produce any rabbits. Since we have learned to experiment with one factor at a time we now take the doe out and replace it with a buck. After waiting for some time things begin to look black and we might conclude that it is simply not possible to breed rabbits. We have now tried all the combinations in Figure 9.6 except the upper right one. Adding this last case, *combining* a doe and a buck, we do not only produce a full factorial experiment but a number of little rabbits as well!

The important message is that many important phenomena depend on a combination of factors; "And (I'm thankful to say) you can't get rid of this phenomenon by transformation", Box adds [2]. Interactions do not occur only in biology. Investigating the conversion efficiency of a catalyst, for example, you will often find that it works best at a certain combination of reactant concentrations. Box shows how the life of a rolling bearing may be increased fivefold by successful combinations of factors [2]. When baking cookies we rely on a correct combination of oven temperature and baking time – if one changes, the other must change too. When we think about it, the world is full of situations where a system relies on several factors coming together at the appropriate settings. Such factor interactions are difficult to detect when experimenting with one factor at a time and this is an important motivation for using design of experiments.

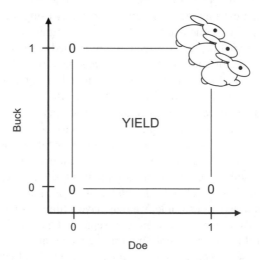

Figure 9.6
Imaginary rabbit breeding experiment.

9.5   Factor Screening: Fractional Factorial Designs

At the beginning of a study it is often possible to think of a large number of factors that could be important for the response. Initially, we may only be interested in finding out which of these candidates – if any – have a substantial effect. This process of separating the vital few factors from the trivial many is called *screening*. The purpose is to find out which factors to keep for the real experiment, which is to be conducted later.

Equation 9.2 shows that the number of runs in a full factorial experiment increases exponentially with the number of factors. This is why it is convenient to use special screening designs to reduce the number of runs. We will soon see that this is done by "mixing effects together" and focusing on main effects and low order interactions. These designs typically fall into two categories: *fractional factorial designs* and *Plackett–Burman designs*. We will concentrate on the first category and only briefly mention some characteristics of the latter.

> **Exercise 9.5:** Use Equation 9.2 to determine the number of runs needed in a full factorial experiment as the number of factors, n, increases from two to eight. The number of levels, l, should be two. What can be said about how the number of runs increases when factors are added?

The 2^3 design in Table 9.3 tests three factors in eight runs (replicates have been left out for simplicity). Imagine that you are interested in four factors but are only allowed eight runs. What would you do? The whole matrix corresponds to the complete information set for a 2^3 design, so the fourth factor must come from one of these columns. Our rule of thumb tells us that the highest order interactions are probably not carrying any information when n increases beyond three. This means that the three-way interaction ABC is least likely to carry any information. We could "cheat" and simply call the ABC column D, and vary our fourth factor according to it. We obtain a design that tests the influence of four factors in only eight runs – half of what would be needed in a true full factorial design. It was obtained by *confounding* or *aliasing* the factor D with ABC. We will now briefly discuss some implications of this.

First of all we should note that the interaction ABC *could* have an effect on the response, even though it is not very likely. If so, we cannot separate this effect from the effect of D since they are completely mixed up with each other. In mathematical terms, ABC and D are perfectly correlated. But this is only the beginning. Table 9.5 shows the set of interactions for the new design, rearranged to highlight additional confounding. We see that not only is D confounded with ABC but that each main effect is confounded with a three-way interaction. We also see that the AB column is identical to the CD column. The same is true of AC and BD, and of BC and AD. When confounding D with ABC we have introduced confounding in all other terms as well. This means that the design is not orthogonal anymore – all effects are blended with other effects.

By confounding D with ABC we have generated a design that allows us to test the influence of four factors with half the number of runs that would be required for a full factorial design. The design is therefore called a *half factorial design*. It is also referred to as a 2^{4-1} design, since it is a two-level design testing four factors in $2^{4-1} = 8$ runs. Reducing the number of runs even more we would get a *quarter factorial design* (2^{4-2}). These types of reduced designs are generally referred to as *fractional factorial designs*. The test matrix

Table 9.5 Two-level half factorial design with four factors. The columns have been rearranged to highlight the confounding.

A	BCD	B	ACD	C	ABD	D	ABC
−1	−1	−1	−1	−1	−1	−1	−1
1	1	−1	−1	−1	−1	1	1
−1	−1	1	1	−1	−1	1	1
1	1	1	1	−1	−1	−1	−1
−1	−1	−1	−1	1	1	1	1
1	1	−1	−1	1	1	−1	−1
−1	−1	1	1	1	1	−1	−1
1	1	1	1	1	1	1	1

AB	CD	AC	BD	AD	BC
1	1	1	1	1	1
−1	−1	−1	−1	1	1
−1	−1	1	1	−1	−1
1	1	−1	−1	−1	−1
1	1	−1	−1	−1	−1
−1	−1	1	1	−1	−1
−1	−1	−1	−1	1	1
1	1	1	1	1	1

in Table 9.5 is analyzed in exactly the same way as the full factorial design, obtaining effects by taking the inner products of the columns and the response. The big difference is that the interactions are mixed up with each other. If *AB* seems to be significant, we do not know if *AB* or *CD* is responsible for the effect. We simply pay for reducing the number of runs by obtaining less detailed information.

In addition to reducing the size of an experiment, confounding is a suitable method for introducing a block variable. This is because the confounding disappears if the block variable turns out to be unimportant. The experiment then provides more detailed information about the remaining factors.

9.6 Determining the Confounding Pattern

Before proceeding with the fractional factorial designs, we need a method to determine how the factors and interactions are confounded. Without this knowledge it is not possible to interpret the results. One way to do it is to use what is called the *defining relation* of the design.

Firstly, we may note that multiplying one of the columns in the matrix with itself yields a column where every element equals one. (In vector multiplication each element of one vector is multiplied with the corresponding element in the other.) A column of ones is called an identity vector and is denoted *I*. It gets its name from the fact that multiplying a vector with *I* will return the vector itself. When *D* is confounded with *ABC* we get *ABCD* = *DD* = I, which is the defining relationship of our design:

$$I = ABCD \tag{9.5}$$

This relationship can be used to identify the confounding pattern. Multiplying a column in the matrix by $ABCD$ returns the column that it is confounded with. For example, A times $ABCD$ equals $IBCD$, which is equivalent to BCD. The defining relation also gives that A times $ABCD$ equals AI, or A. This means that A is confounded with BCD.

We could do the same thing with an interaction column; for example, $AB = ABI$. According to the defining relation this is equivalent to $ABABCD = AABBCD = IICD = CD$, meaning that AB is confounded with CD.

To turn these calculations into a simple rule, the confounding of a factor or interaction is obtained by simply removing the letters designating that factor or interaction from the defining relation. Removing A from $ABCD$ leaves us with BCD, and removing AB leaves CD.

> **Exercise 9.6:** Use this rule to obtain the complete confounding pattern of the design.

Another way to determine the confounding pattern (or degree of orthogonality) in a design is to multiply the matrix containing the factors and interactions by itself. We will call this matrix **X**. To do this multiplication in Excel the matrix must first be transposed, meaning that the rows are turned into columns and vice versa. The transpose of **X** is called **X′** and the multiplication is shown in Table 9.6. Only terms up to second-order interactions are shown for clarity. Note that each entry in **X′X** is obtained by taking the inner product between a row in **X′** and a column in **X**. For example, the bottom right element in the ten-by-ten matrix **X′X** is the inner product between the tenth row of **X′** and the tenth

Table 9.6 Orthogonality check of the matrix **X**, which contains the factors and interactions up to second order. If **X** were orthogonal, only the diagonal of **X′X** would contain non-zero elements.

X′

A	−1	1	−1	1	−1	1	−1	1
B	−1	−1	1	1	−1	−1	1	1
C	−1	−1	−1	−1	1	1	1	1
D	−1	1	1	−1	1	−1	−1	1
AB	1	−1	−1	1	1	−1	−1	1
AC	1	−1	1	−1	−1	1	−1	1
AD	1	1	−1	−1	−1	−1	1	1
BC	1	1	−1	−1	−1	−1	1	1
BD	1	−1	1	−1	−1	1	−1	1
CD	1	−1	−1	1	1	−1	−1	1

X

A	B	C	D	AB	AC	AD	BC	BD	CD
−1	−1	−1	−1	1	1	1	1	1	1
1	−1	−1	1	−1	−1	1	1	−1	−1
−1	1	−1	1	−1	1	−1	−1	1	−1
1	1	−1	−1	1	−1	−1	−1	−1	1
−1	−1	1	1	1	−1	−1	−1	−1	1
1	−1	1	−1	−1	1	−1	−1	1	−1
−1	1	1	−1	−1	−1	1	1	−1	−1
1	1	1	1	1	1	1	1	1	1

X′X

	A	B	C	D	AB	AC	AD	BC	BD	CD
A	8	0	0	0	0	0	0	0	0	0
B	0	8	0	0	0	0	0	0	0	0
C	0	0	8	0	0	0	0	0	0	0
D	0	0	0	8	0	0	0	0	0	0
AB	0	0	0	0	8	0	0	0	0	8
AC	0	0	0	0	0	8	0	0	8	0
AD	0	0	0	0	0	0	8	8	0	0
BC	0	0	0	0	0	0	8	8	0	0
BD	0	0	0	0	0	8	0	0	8	0
CD	0	0	0	0	8	0	0	0	0	8

column of **X**. This sort of multiplication can be confusing if you are not used to matrix operations but it is easily performed in Excel using the MMULT function. The only difficulty involved is finding out how to enter the MMULT expression into the worksheet as an array formula.

The important thing to remember is that, if **X** is orthogonal, **X'X** will only contain non-zero elements on the diagonal. This is because the off-diagonal elements contain the inner products between separate columns in **X**, whereas the diagonal elements result from multiplying the columns with themselves. Looking at **X'X** in Table 9.6 we see that it is not completely orthogonal, since a few off-diagonal elements have non-zero values. Looking closer, we see that these elements correspond to the confounded two-way interactions $AB = CD$, $AC = BD$, and $AD = BC$. This means that we have obtained the complete confounding pattern of the design in a single matrix operation. The larger the design, the more convenient this method becomes.

Exercise 9.7: Use Excel to make the orthogonality check in Table 9.6. Firstly, enter the matrix **X** into the worksheet. Transpose it by copying the cells and choosing "Paste Special" in the Edit menu. In the dialog that appears, choose "Paste Values" and check the "Transpose" box. This returns the matrix **X'**, which can now be multiplied by **X**. Do this by writing "=MMULT(**X'**;**X**)" into a cell (the matrices should of course be replaced with their cell references, for example "A1:J8"). Pressing enter will return the first element of **X'X**. To get the full matrix, select a ten-by-ten area with the first element in the upper left corner. After this, press F2 and then CTRL+SHIFT+ENTER. (In Mac OS X this corresponds to CTRL+U followed by CMD+RETURN.) You should now have the full **X'X** matrix as shown in Table 9.6.

9.7 Design Resolution

The number of factors occurring in the defining relationship is called the *resolution* and describes the degree of confounding in a fractional factorial design. The higher the resolution, the more information is obtained from the experiment. The resolution is written using roman numerals. For example, if the defining relation is $I = ABCD$ the resolution is IV (four). It is easy to visualize the types of confounding that occur with a resolution of IV by holding up four fingers on one hand. Figure 9.7 shows that you can group them two and two (index and middle finger in one group, ring and little finger in the other). This represents confounding between a pair of two-way interactions. You can also group them as one plus three (for instance index finger in one group and the rest in the other). This represents confounding between main effects and three-way interactions.

With a resolution V design you could either group the five fingers as one plus four or as two plus three. This represents confounding between main effects and four-way interactions, and between two- and three-way interactions, respectively.

Exercise 9.8: Use your fingers to determine the degree of confounding of a design with resolution III.

Figure 9.7
A simple technique for translating resolution into confounding patterns, here shown for resolution IV. Grouping four fingers two and two represents confounding between a pair of two-way interactions. Grouping them one and three corresponds to confounding between a main effect and a three-way interaction. © Öivind Andersson.

Using this method it is easy to see that resolution II designs are not useful, since they confound main effects with other main effects. Similarly, designs with a resolution above V may be considered wasteful as they estimate interactions of such high order that they are unlikely to occur. Resolutions from III to V are most frequent. Table 9.7 gives the resolutions of fractional factorial designs for given numbers of factors and runs. Those who run fractional factorial experiments will probably find this table helpful for choosing an appropriate design. Note that when five or more factors are used, there are fractional factorial designs that can be preferred over full factorials as they give sufficiently detailed information in fewer runs. Fractional designs, in other words, are not only useful for screening.

9.8 Working with Screening Designs

An interesting property of the half factorial design is that, if one factor turns out to be unimportant, the design collapses to a full factorial design. This is most easily understood

Table 9.7 Resolutions for fractional factorial designs with various numbers of factors and treatments.

	FACTORS									
TREATMENTS	1	2	3	4	5	6	7	8	9	10
4		Full	III							
8			Full	IV	III	III	III			
16				Full	V	IV	IV	IV	III	III
32					Full	VI	IV	IV	IV	IV
64						Full	VII	V	IV	IV
128							Full	VIII	VI	V

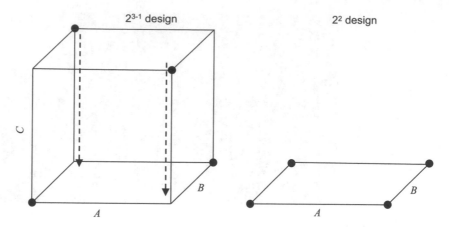

Figure 9.8
The three-factor half factorial design may collapse to a two-factor full factorial design if one factor (here C) turns out to be unimportant.

looking at it graphically. The left part of Figure 9.8 shows a three-factor half factorial design. The dots in the corners show the treatments used and we see that every other treatment is missing from the corresponding full factorial design, which is represented by all the corners of the cube. If, for instance, factor C turns out not to have a significant effect on the response, this means that there is no discernible difference in the response when C is switched between its two levels. In other words, a change in the setting of C only measures the experimental error. This could be seen as the cube collapsing along the C-direction. As shown in the right-hand part of the figure, we then obtain a *full factorial design* with the remaining two factors A and B. If these factors turn out to be significant, we now have data for a more detailed study of them.

The fact that every other corner point is missing can be advantageous in other ways than just reducing the number of runs. In the corners, all factors are set to their extreme settings. In some cases it is not safe or even possible to run experiments at such extreme conditions. If we are worried about particular combinations of settings, these reduced designs allow us to choose a fraction of the runs that avoids these settings.

Now, let us run a real experiment to see how screening and analysis work in factorial experiments. You will design a paper helicopter, optimizing the settings of five factors in no more than 16 runs. The experiment should normally include at least one replicate but it is omitted in this example to limit your workload. The helicopter blueprint is shown in Figure 9.9. You should take photocopies and cut helicopters from them with a pair of scissors. The factors to be screened are the tail length, tail width, wing length, wing width and the weight. To keep the weight constant when varying the other factors you should cut along the solid lines and vary the factors by folding the paper along the appropriate dashed lines. This keeps the total amount of paper in the helicopter constant. The weight is manipulated by attaching a paper clip to the bottom of the body. Fold the wings in separate directions at the dotted line to obtain a "T"-shaped paper figure, as shown in the insert to the left of the blueprint. The response variable to be maximized is the flight time, so you will need a stopwatch. To fly the helicopter, hold the body between your thumb and index finger and drop it from a predetermined height.

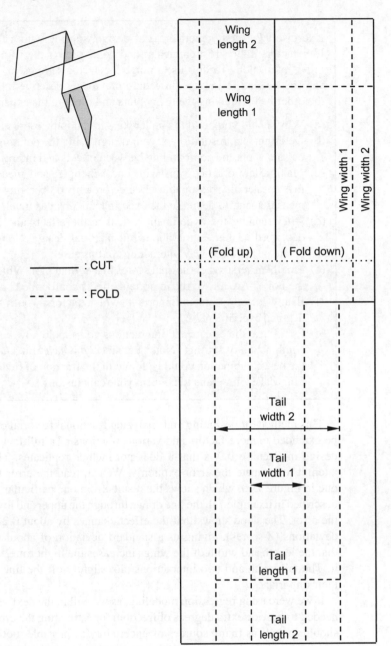

Figure 9.9
Blueprint for the helicopter experiment. Take photocopies, cut along solid lines and fold along dashed ones.

> **Exercise 9.9:** First, generate a 2^{5-1} design, starting with a full factorial design with four factors ($2^4 = 16$ treatments) and confounding the fifth factor with the highest order interaction. To do this, arrange the factor settings in Yates order in an Excel spreadsheet and follow the procedure previously described. Remember to randomize the order before making the helicopters and running the experiment:
>
> (a) Why should you build the helicopters in random order?
> (b) Determine the resolution of your design using the defining relationship. What can be said about the confounding between effects and interactions?
> (c) Make a few practice runs with one helicopter and measure the flight times. Is the repeatability reasonable? Decide on a fixed operating procedure for flight and measurement, to decrease the experimental error as much as possible.
> (d) After running the experiment, calculate the effects and two-way interactions as described in the section about full factorial designs. Why is it not necessary to calculate the three-way interactions in this case?
> (e) Plot them versus their normal scores in a normal plot. Which effects stand out from the noise? Are any of the important effects confounded?
> (f) If any factors seem unimportant, which factors would you choose for a more detailed experiment?
> (g) Looking at the effects and interactions you should now be able to find the optimal combination of settings. Note that, since you have run a *half* factorial experiment, the best combination could be found in the fraction of treatments that you have not run. Make a few runs to confirm your conclusions.

This method for designing and analyzing fractional factorial experiments can, of course, be extended to more factors and various fractions of a full factorial design. A lack in the design in Exercise 9.9 is that it does not include replicates. This is because we did not attempt to estimate the error explicitly. We can read the error from a normal plot as the one in Figure 9.10, which shows the results from one particular helicopter experiment. As described in Example 7.4, the line drawn through the upper and lower quartiles approximates the error. The slope shows that the effect changes by about 0.2 seconds in three standard deviations (Z-scores), giving us a standard deviation of about 0.07 s. The plot indicates that the length and width of the wings increases the flight time, whereas the clip decreases it. The tail width and two interactions plot slightly off the line but it is difficult to judge their importance.

If we were using regression modeling, as we will in the next section, replicates would be needed to provide extra degrees of freedom for estimating the error. Note that the replicates should include as many sources of uncertainty as possible, both from the "manufacture" and flying of the helicopter. With one replicate run this would require us to make two helicopters for each treatment.

Despite the lack of replicates, two important precautions were employed in this experiment: randomization and a fixed operating procedure. Needless to say, it is important to settle on a fixed drop height, a fixed method of releasing the helicopter, a predefined manner of counting down to the drop, appointing a single person to operate the timer, and so forth. To obtain a feeling for how shortcomings in the experimental procedure can influence the result it is useful to carry the exercise out in several teams and compare the outcomes. It is, for example, common to find that different teams have made the helicopters in slightly different ways, since even a detailed building instruction leaves room for interpretation.

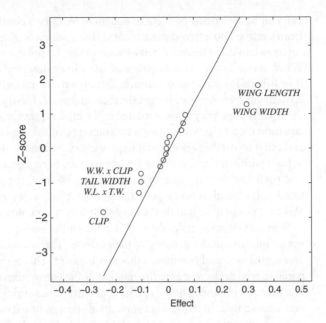

Figure 9.10
Normal plot from a helicopter experiment.

Make sure to fly the best helicopters from different teams using *one* operating procedure and then discuss how any differences may be explained. This exercise demonstrates the need for describing experimental procedures in great detail when publishing research results. The reader must be able to repeat both the experiment and the outcome.

Before ending this section we will briefly mention another class of screening designs, called Plackett–Burman designs. As we have seen, the number of treatments in fractional factorial designs is always a power of two: 4, 8, 16, 32, 64 and so on. The gaps between the numbers increase when the number of factors increases. With a very large number of factors it may be difficult to find a design of reasonable size. With Plackett–Burman designs the number of treatments is a multiple of four. For example, designs with 12, 20, 24 and 36 treatments can be obtained. This makes it easier to find a design that matches your experimental limitations. Just as factorial designs, Plackett–Burman designs are orthogonal matrices whose elements are all +1 or −1, but they are smaller. Technically, they are Hadamard matrices and you can find free libraries of them by searching the internet for "Plackett–Burman designs" or "Hadamard matrices". Plackett–Burman experiments are analyzed analogously to factorial experiments. A drawback of these designs is that the main effects are confounded with two-way interactions. This means that, if there are important interactions in the system under study, these will be mistaken for main effects in the analysis. Considering that screening experiments are made because we do not know which factors and interactions are important, it is rarely a safe assumption that interactions do not exist. For this reason, Plackett–Burman designs should be applied with caution.

9.9 Continuous Factors: Regression and Response Surface Methods

In many important cases the factors of an experiment are not categorical but continuous, numerical variables. When these are varied the result is often a smoothly varying response that can be described by a curve equation. With two continuous factors the response turns from a curve into a two-dimensional surface, which is often called a *response surface*. Such surfaces can often be described by second-degree polynomials, as the one in Equation 8.13. When more factors are investigated the dimensionality of the polynomial increases but it is still called a response surface. Fitting surface models to experimental data has many advantages over simply plotting the measurements. Firstly, the regressors, or coefficients, of the equation are very closely related to the effects previously introduced. Just looking at the equation thereby gives a quantitative measure of the importance of the factors, interactions and other terms that appear in it as predictors. Secondly, the equation is a predictive model, at least within the range where the factors have been varied. This means that we can analyze and optimize the response mathematically. We can, for instance, take the derivative of a fitted polynomial to find a potential optimum just as we can with any other equation. This makes it possible to handle data even from very complex systems analytically.

When a system is under the simultaneous influence of several factors it is often difficult to get an intuitive understanding of their effects. The combination of orthogonal test matrices and regression analysis makes this much easier. This approach is also very efficient compared to trial-and-error experimentation or the conventional method of varying one variable at a time. Considering how expensive and time consuming it can be to run experiments it is odd to note that, in many important areas, design of experiments is never applied. To give just one example, I once heard the British statistician Tony Greenfield mention a discussion about superconductors in a conference speech [3]. Superconductivity usually occurs at temperatures below 100 K or so, but certain combinations of materials are capable of raising the temperature limit. Researchers believe that a specific combination of seven or eight metals may reach up to the magic 293 K. Since room temperature superconductivity would allow massive energy savings across the world, this would be a momentous breakthrough. Greenfield was surprised to hear that the researchers tried to find the optimum combination by varying one parameter at a time. When he met the director of the superconductivity research laboratory at an internationally renowned university, he suggested that they might discuss the design of experiments. The scientist replied that it would not work because "in physics research, we have everything under control. That means we must study only one variable at a time to reach a full understanding of its behavior. And that is why it will take a very long time." In his speech, Greenfield commented "His logic escaped me" [3].

When varying one factor while holding all the others constant we assume that there are no interactions between them. We bluntly assume that the effect of one factor is simply added to that of another. Figure 9.11 illustrates how dangerous this assumption can be. The contour plot shows a response to variation in two variables x and y, which has a maximum in the middle of the plot. Imagine that we do not know what this response surface looks like. Setting out to find the maximum by varying one factor at a time, we may start by fixing y at 19 and taking measurements at the x settings marked by plus signs in the diagram. This will give us the impression that the optimal setting of x is 19.5. Now, keeping x fixed at this value we continue by sweeping y as indicated by the circles in the vertical direction. We find that the response can be increased a little if y is also set to 19.5, but looking at the contours this is obviously not the maximum. What is worse is that if we had made the initial sweep in x at another setting of y – say, at 20.5 – we would have ended up at another "maximum". The end result, in other words, depends on the starting point.

You might argue that you would not stop at this point but continue by iteratively sweeping x and y until you felt that you were close to the peak. It would still require luck to find the true optimum but the important point is that it can be done more economically and most

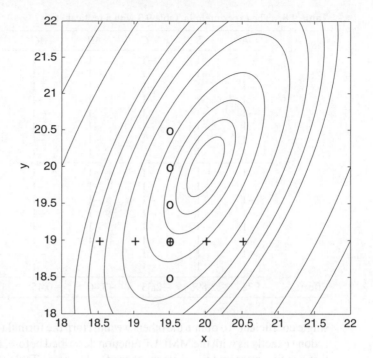

Figure 9.11
Response surface based on nine measurements. Two single-variable sweeps are represented by plus signs and circles. The response surface provides more complete information than the sweeps and reveals the interaction between the two variables x and y.

likely with greater precision using a response surface method; that is, by combining design of experiments and regression modeling. The contours in Figure 9.11 are based on nine measurements and, apart from showing the true optimum, they give much more information about the response than the nine measurements in the two sweeps.

We will start by making a regression model of the data in Table 9.8. It contains the same data as the full factorial experiment in Table 9.3 but a replicate run has been added in the lower half. Due to the experimental error the replicated responses are a little different from the original ones, which is why the calculated effects differ slightly from those in Table 9.3. We will, for the moment, pretend that the factors are numerical instead of categorical; otherwise a regression model would make no sense. The regression model is easily obtained by the Excel worksheet function LINEST, which provides least squares estimates of the coefficients. The first input argument to this function is the response column. The second input is the range of columns corresponding to the factors and interactions. We also need to specify if the constant term in the polynomial should be zero or not; that is, if the response surface is to be forced to go through the origin. Finally, we can request the function to calculate regression statistics.

After entering Table 9.8 into an Excel sheet, the first step is to write "=LINEST($\mathbf{Y}$, $\mathbf{X}$, TRUE, TRUE)" into a cell, where $\mathbf{Y}$ should be replaced with the cell references for the response column and $\mathbf{X}$ with the references of the seven columns from A to ABC. The first "TRUE" means that the constant is to be calculated normally and not forced to be zero. The second "TRUE" makes the function return regression statistics. Pressing enter will return a

Experimental Design

Table 9.8   The experiment in Table 9.3 with a replicate run.

A	B	C	AB	AC	BC	ABC	RESPONSE
−1	−1	−1	1	1	1	−1	56.2
1	−1	−1	−1	−1	1	1	55.6
−1	1	−1	−1	1	−1	1	61.6
1	1	−1	1	−1	−1	−1	52.2
−1	−1	1	1	−1	−1	1	54.0
1	−1	1	−1	1	−1	−1	50.0
−1	1	1	−1	−1	1	−1	60.3
1	1	1	1	1	1	1	51.1
−1	−1	−1	1	1	1	−1	54.8
1	−1	−1	−1	−1	1	1	54.2
−1	1	−1	−1	1	−1	1	61.6
1	1	−1	1	−1	−1	−1	50.9
−1	−1	1	1	−1	−1	1	52.9
1	−1	1	−1	1	−1	−1	48.6
−1	1	1	−1	−1	1	−1	59.5
1	1	1	1	1	1	1	52.1
Effect: −5.78	2.89	−2.33	−3.40	−0.45	1.50	1.31	

single coefficient. To obtain the others we must turn the formula into an array formula, which is done exactly as with the MMULT function described before. Firstly, select a range of cells that has five rows and the same number of columns as Table 9.8. The LINEST expression should be in the top left cell. After this, press F2 and then CTRL+SHIFT+ENTER (or, in Mac OS X, CTRL+U followed by CMD+RETURN). We now obtain the numbers in Table 9.9, which will be explained presently.

> **Exercise 9.10:** Before continuing, carry out the above operations in Excel to obtain the numbers in Table 9.9.

The numbers between the lines in Table 9.9 are the output from LINEST. Cells that are relevant to this example have been shaded. The first row contains the coefficients of the polynomial, but as seen in the header they are presented in reverse order to the columns in Table 9.8. If you look carefully you will find that the coefficients are exactly half the effect values calculated in Table 9.8. Recall that an effect is the average change in the response when the factor is switched from −1 to +1. Since the coefficients are slopes, they

Table 9.9   Output from the LINEST function; model coefficient and regression statistics.

	ABC	BC	AC	AB	C	B	A	CONSTANT
Coefficient:	0.66	0.75	−0.22	−1.70	−1.17	1.45	−2.89	54.73
Standard error:	0.200	0.200	0.200	0.200	0.200	0.200	0.200	0.200
R-squared:	0.980	0.801	#N/A	#N/A	#N/A	#N/A	#N/A	#N/A
v:	56.03	8	#N/A	#N/A	#N/A	#N/A	#N/A	#N/A
	251.741	5.135	#N/A	#N/A	#N/A	#N/A	#N/A	#N/A
t:	3.28	3.75	1.12	8.49	5.82	7.22	14.42	273.27
p:	0.011	0.006	0.296	0.000	0.000	0.000	0.000	0.000

measure the change in response *per unit change* in the factor. As each factor changes by two units, the coefficients are the effects divided by two. The constant is just the mean of all responses, according to the principles explained in connection with Figure 8.11. Using these coefficients the response, Y, can now be described as:

$$Y = 54.73 - 2.89A + 1.45B - 1.17C - 1.7AB - 0.22AC + 0.75BC + 0.66ABC$$

$$(9.6)$$

The following are the most important parts of the rest of Table 9.9. The second row of the table contains the standard errors of the coefficients. These are measures of the uncertainty in the coefficient estimates. The shaded cell on the third row shows R^2 (the coefficient of determination), which was discussed in Chapter 8. It shows that the model is quite good, accounting for 98% of the variation in the data. Finally, we find the number of degrees of freedom, v, on the fourth row. If you are interested in the other regression statistics, they are explained in the help section about LINEST in Excel.

We are now able to calculate t-statistics for the coefficients to estimate how useful they are in predicting the response. The observed t-value is simply the coefficient divided by its standard error. It is given for each coefficient below the table. These, in turn, can be used to calculate p-values using the TDIST function. As explained in the last chapter, writing "=TDIST(t;v;2)" into a cell returns the p-value, which is the probability of obtaining a coefficient of at least this magnitude under the assumption of the null hypothesis that the factor has no effect on the response. In the expression, t is the observed t-value of the coefficient, v the number of degrees of freedom, and the last entry makes it a two-tailed test. The p-values are given at the bottom of the table where only the AC coefficient has $p > 0.05$. Applying a significance level of 5% we would consider the others to be statistically significant.

> **Exercise 9.11:** Calculate the responses predicted by the model for each treatment by applying Equation 9.6 to a new column in the Excel worksheet. Then calculate the residuals and determine R^2 using Equation 8.12.

It is crucial to note that a two-level factorial design is only appropriate for categorical factors. With continuous, numerical factors, the design in Table 9.8 suffers from a serious drawback, since a design that only tests two settings for each factor is incapable of detecting curvature. Any non-linearity in the response will thus go unnoticed. This example illustrated how effects and coefficients are related but to detect curvature we need more information. One method is to add a treatment where all factors are centered in the test matrix, a so-called *center point*. If there is curvature, the response in the center point will deviate from the constant term in the model.

However, a center point is not sufficient for fitting a polynomial with quadratic terms to the data. This is because estimating the coefficients of these terms requires several additional degrees of freedom. In general, we need at least as many treatments as the number of coefficients that we wish to estimate. A three-level factorial design could be used but it actually provides unnecessary many degrees of freedom. It is more efficient to add axial points to the two-level design. Figure 9.12 shows an example of such a design. Axes are drawn through the center of the cube and the axial points are, in this case, located at the faces of the cube. This type of design is a composite of a two-level factorial design (full or fractional), the center point and the axial points. For this reason it is called a *central composite design*. It provides full support for a quadratic model while using fewer

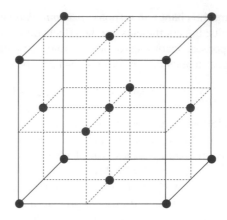

Figure 9.12
Central composite design with three factors. The axial points here are centered on the faces of the cube.

treatments than a three-level factorial design. Table 9.10 clearly shows how economic this design is compared to a three-level full factorial design, especially when the number of factors increases. Comparing Figure 9.12 to the three-level full factorial design in Figure 9.3 we see that they are similar but that the points on the edges of the cube are missing.

The axial points could also be located at other distances; for example, so that all points in the design are at equal distances from the center point. This produces a *rotable* design. The advantage of this is that, for reasons beyond the scope of this book, the precision of the model will depend only on the distance to the origin, giving equal precision in all coefficient estimates. A potential drawback is that the factors must be set to five levels instead of three, which may make the experiment more difficult to run.

Since central composite designs contain more treatments than two-level designs, replicates become more expensive. For this reason, it is common to replicate only the center point but to do it several times. Apart from providing the necessary degrees of freedom for estimating the experimental error, these replicates can be used to detect drift in the experimental conditions if they are well distributed in time. This is especially useful in cases when it is difficult or even impossible to use a fully randomized run order. We should note that using center points to estimate the error rests on the rather optimistic assumption that the error is constant over the entire range of the factors, which is frequently not true. Many systems become unstable when several factors are set to extreme values in the corner points; this will increase the error locally.

To proceed with analyzing a central composite experiment, Table 9.11 shows data from an experiment on a diesel engine. The measured response is the level of nitrogen oxides (NO_x) in the exhausts, which is to be reduced. Three factors are of interest. NO_x is often

Table 9.10   Number of treatments for central composite designs and the corresponding three-level full factorial designs. The number of model coefficients to be estimated is also given for various numbers of factors.

Factors:	3	4	5	6
Coefficients (second-degree model):	10	15	21	28
Treatments (central composite):	15	25	27	45
Treatments (3-level full factorial):	27	81	243	729

Table 9.11 Central composite design applied to a diesel engine.

O	P	D	$NO_x(1)$	$NO_x(2)$
-1	-1	-1	0.4	0.6
1	-1	-1	7	7.1
-1	1	-1	0.5	0.5
1	1	-1	7.5	7.7
-1	-1	1	0.5	0.5
1	-1	1	7.2	7.2
-1	1	1	0.5	0.5
1	1	1	7.2	7.3
0	0	0	2	2.2
-1	0	0	0.6	0.7
1	0	0	8.1	8.1
0	-1	0	1.9	2.1
0	1	0	1.7	1.8
0	0	-1	1.8	1.8
0	0	1	1.9	1.9

	PHYSICAL UNITS:		
CODED UNITS:	O (%)	P (bar)	D (kg/m^3)
-1	12	1500	24
0	14	2000	27
1	16	2500	30

reduced by diluting the intake air with exhaust gases. This reduces the oxygen concentration in the intake, represented by the factor "O" in the table. The other factors are the fuel injection pressure ("P") and the gas density in the cylinder after compression by the piston ("D"). The second response column contains replicates of the entire test plan. Although the measurements were randomized, they are presented here in standard order for clarity. The upper part of the table is the familiar 2^3 full factorial design, followed by the center point $(0, 0, 0)$ and the axial points. The latter are easily identified as all factors except one are set to zero on each row. Beside the test matrix there is a table showing the factor settings in physical units. There are two reasons for using coded units with the settings -1, 0 and 1 in the test matrix. Firstly, they make it is easier to see the structure of the design. Secondly, it facilitates comparison of the effects. This is because the magnitudes of the coefficients in the polynomial will depend on the units of the predictor variables.

To illustrate why this is so, imagine that you are baking cookies and want to determine whether the baking temperature or the baking time is more important for the result. Say that you measure the temperature in Kelvin. Depending on whether you measure the time in seconds or hours, its numerical value will vary by four orders of magnitude. As the estimated coefficient for time is a slope, it too will vary by four orders of magnitude. The coefficient for the temperature may therefore be either smaller or greater than the coefficient for time, depending on which unit you use. This makes it difficult to estimate the relative importance of the two factors. With coded units this difficulty is avoided since the physical units are removed. Instead, the coefficients are determined from non-dimensional, normalized factors.

Now, let us return to our diesel engine. To analyze this experiment in Excel, the replicate runs should be entered below the test plan to obtain a table with 30 rows and a single response column. A column should also be added for each term that is to be included in the model. In this case we are interested in the interactions (OP, OD, PD, and OPD) as well as the quadratic terms (OO, PP and DD). These columns are obtained by multiplying the factor columns with each other as previously described. Using LINEST we obtain the regressors and statistics, and from these we can calculate the p-values just as before. The result is shown in Table 9.12.

Table 9.12 Regression output for the diesel engine data in Table 9.11.

	DD	PP	OO	OPD	PD	OD	OP	D	P	O	CONSTANT
	−0.26	−0.23	2.27	−0.06	−0.06	−0.02	0.07	−0.01	0.04	3.46	2.10
	0.079	0.079	0.079	0.045	0.045	0.045	0.045	0.040	0.040	0.040	0.068
	0.998	0.179	#N/A	#N/A	#N/A	#N/A	#N/A	#N/A	#N/A	#N/A	#N/A
	835.70	19	#N/A	#N/A	#N/A	#N/A	#N/A	#N/A	#N/A	#N/A	#N/A
	268.548	0.611	#N/A	#N/A	#N/A	#N/A	#N/A	#N/A	#N/A	#N/A	#N/A
t:	3.23	2.92	28.71	1.39	1.39	0.56	1.67	0.25	0.87	86.19	30.89
p:	0.004	0.009	0.000	0.179	0.179	0.583	0.111	0.806	0.393	0.000	0.000

> **Exercise 9.12:** Enter the data into a worksheet as described above and generate the regression model and statistics using LINEST.

The R^2-value shows that the model is good, accounting for 99.8% of the variation in the data. Looking at the p-values, however, we see that a number of coefficients are small in comparison with the noise level (shaded cells). When discussing regression in Chapter 8 we said that removing such terms would give us a more parsimonious model that more clearly illustrates the relationships in the data. However, removing the "insignificant" D and P factors would also remove the higher order terms where they appear. Since all the quadratic terms are important, this would not result in a useful model. If we want to trim the model we could try to remove the interaction terms and see if the model remains useful.

Looking at the coefficients we see that the oxygen concentration has the largest effect on NO_x, and that there is a strong curvature in the response to this factor. There are no apparent interactions between the factors, making the interpretation more straightforward. It is useful to make a Pareto plot of the coefficients to obtain a graphical overview of the effect sizes. Figure 9.13 shows such a plot, where bars sorted in declining order represent

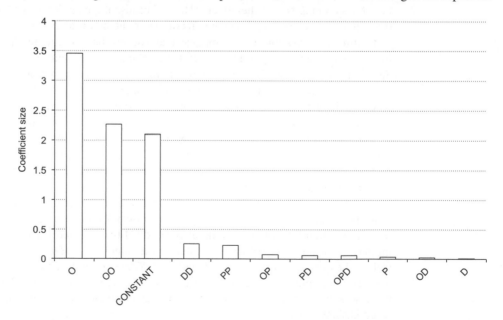

Figure 9.13
Pareto plot of the model terms obtained from the diesel engine experiment.

the absolute values of the coefficients. Clearly, the oxygen terms represent nearly all the variation in the response, followed by the quadratic terms for density and injection pressure.

Having obtained the model, we are, of course, interested in how well it predicts the response. Making prediction calculations is quite straightforward in Excel, albeit a bit laborious. It is easiest to copy the coefficients from the LINEST output and paste them into a separate worksheet. Make sure to use "Paste Special" and select "Paste Values", and to label the coefficients so you know which term they belong to. Then make three factor columns, labeled O, P and D, and an output column. Enter the input values you are interested in into the factor columns. In the output column, you enter the formula for the fitted polynomial. This is done as follows:

Start by typing an equals sign into the first cell of the output column. Then add the terms of the polynomial one by one. The first term is the cell reference for the constant, which is obtained by simply clicking that cell. Add a plus sign followed by the second term, which is the cell reference for the O coefficient multiplied by the cell reference for the first cell in the O factor column. Continue in the same manner until you have all the terms in place. Now, lock the references for the *coefficients* by typing a "$" sign before the letter and number of their cell references. For example, if the constant is in cell C3, the cell reference should read "\$C\$3". After this, mark the first output cell and double-click the fill handle to apply the calculation to all the cells of the output column.

As a first step, it is useful to enter the test matrix into the factor columns. The output column will then contain the modeled predictions of all the measured values. If we plot these values versus the observed ones we obtain the diagram in Figure 9.14. The model is obviously good since the points fall on a straight line with a slope of one. The residuals are small and are randomly scattered about the line. If the points plot off the straight line, you could have made a mistake when entering the formula into the spreadsheet. It could also be due to a poor model, but that would be reflected in a low R^2 value.

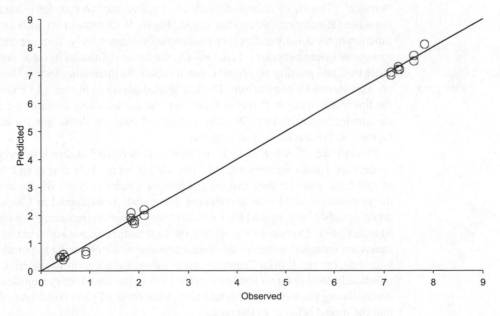

Figure 9.14
Observed responses plotted versus the predicted ones utilizing data from the diesel engine experiment.

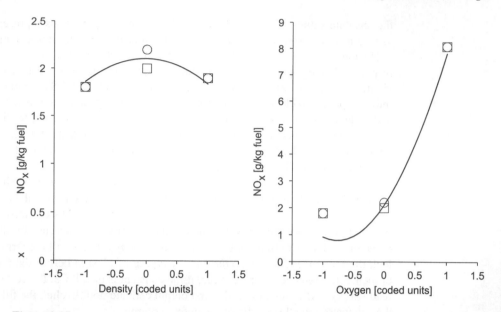

Figure 9.15
Modeled NO_x response as function of density and oxygen. Measured values are shown as a reference.

Exercise 9.13: Calculate the predicted values corresponding to the measurements as described above, and make an "observed versus predicted" plot. The result should look like Figure 9.14.

Since the model contains three input variables, the response surface is a three-dimensional "surface". This is, of course, difficult to visualize but we can, for instance, display one-dimensional sections through the origin. Figure 9.15 shows two such curves, of NO_x as function of the density and oxygen concentration, respectively. They are obtained by putting a range of values between -1 and $+1$ into the factor column of interest, putting zeros in the other two, and plotting the output column versus the interesting factor. The observed values are also shown for comparison. The left-hand diagram in Figure 9.15 shows that, although the linear density term D was insignificant, the model nicely captures the curvature due to the quadratic density term DD. The right-hand diagram shows that the dominant oxygen factor also produces a curved response.

Though the R^2 value is high the model is a bit off at the low oxygen setting. All models are approximations and this deviation is most likely due to an inherent limitation of quadratic models: they can only represent quadratic types of curvature. In this case the response resembles an exponential curve and, as explained in Chapter 8, the model could possibly be improved by a variable transformation, replacing O with an exponential function of O. Experimenting with variable transformations and trimming the model by removing unimportant terms are standard measures for obtaining the most useful model of your data, but recall from Chapter 8 that p-values are not entirely objective. You should not mindlessly remove terms just because they fall below an arbitrary significance level. When manipulating the model you should also make residual plots (see Figure 8.15) to confirm that the model behaves as intended.

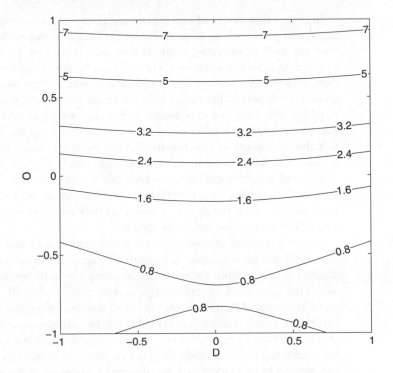

Figure 9.16
Contour plot of the response surface model from the diesel engine experiment showing the combined response to variations in density and oxygen.

It is also possible to show a two-dimensional section through the surface (Figure 9.16). This contour plot shows the combined response to the two factors in Figure 9.15, with the third factor P set to zero. It provides a convenient overview of the response's behavior that contains more information and can be more useful than plots of sweeps of single factors. A drawback is that it is difficult to plot the measured values in contour plots to illustrate the model error.

If we were interested in minimizing the NO_x emissions from the engine we could now do this by taking derivatives with respect to the input variables and finding a zero point, if one exists. If you are working with response surface methods for optimization, however, you will probably use special software for this. In many important cases there are several response variables to be optimized and they may have conflicting dependencies on the factors. It is often impossible to find an unambiguous optimum, as the preferred setting may involve trade-offs between different responses. In such applications, software for *multi-objective optimization* is indispensable. It should also be mentioned that, if optimization is your main objective, it is of course more convenient to use physical units in the model. Otherwise the coded units must be "uncoded" to obtain the preferred setting in useful numbers.

One of the great advantages of central composite designs is that they are suited to sequential experimentation. We may start with a fractional factorial design for screening, using a center point to detect potential curvature. When we have found the interesting factors, we may simply add the axial points to the design and, in just a few extra runs, we obtain sufficient data to build a quadratic response surface model.

In areas where design of experiments and response surface methods are not widely used, prediction plots such as those in Figure 9.15 may be met with some skepticism. People who are used to sweeping single factors may think that three measurement points are insufficient to obtain a precise curve. It should be noted that these curves are just sections through a response surface fitted to 30 measurements. Although the plot only shows three measurement points, the model is based on *all* the measurements. If you are working in a field where design of experiments is not common, your audience may be confused if you present your results in effect plots and prediction plots. The approach still has the valuable advantages of detecting interactions while separating the effects, and of providing a basis for reliable multidimensional response models. If you worry about the skeptics, add a couple of measurement points on each axis through the cube. You can then plot the data as if the factors were swept through the center point. Be sure to keep the corner points of the factorial part of the design, however, as they are an indispensable foundation for the interaction terms of the regression model.

The experimental precautions used in connection with response surface methods are essentially the same as those for factorial designs. There are two main pitfalls involved: discerning effects from the noise and avoiding the influence of background factors. We assess the noise level by replicating measurements – either all of them or only the center point. Background factors are attended to by randomization. Some factors can be difficult to vary and it is often tempting not to randomize the run order of such factors. Non-randomized experiments can be hazardous but, if we are familiar with the system under study and are sure that it behaves in a repeatable manner, they can sometimes be justified. If a potential background factor is identified, its effect can be assessed in the experiment by blocking. It is useful to confirm that there is no drift by plotting replicates and residuals versus time. Another possibility in non-randomized experiments is to use run orders that are orthogonal to the time trend.

A practical drawback of central composite designs is that the corner points test extreme combinations of factor settings. As previously mentioned, a process may become unstable, unsafe, or even impossible to run at such extreme conditions. In those cases, *Box–Behnken designs* are practical alternatives. Apart from the center point they only include points on the edges of the cube (Figure 9.17). Box–Behnken designs thereby avoid setting all factors to extreme settings at once. Another advantage is that they are rotable or nearly rotable

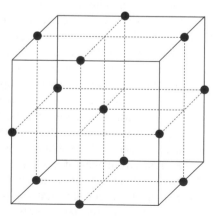

Figure 9.17
Box–Behnken design with three factors.

while only using three factor levels. A drawback is that they are not suited for sequential experimentation, since they use a set of treatments that is complementary to those of the screening designs.

Before finishing the chapter off it should be stressed that designing experiments is one thing and analyzing them is another. The orthogonal designs introduced here help the experimenter to collect data in a way that makes it possible to separate the effects of several factors while also assessing the interactions between them. Regression analysis is a separate activity from the design. It can be applied to any set of continuous data but the regression models become *more reliable* if the test matrix is orthogonal. Any deviation from orthogonality will introduce collinearity between factors and make the model less straightforward to interpret.

Most practitioners of design of experiments work in dedicated software packages. This may be convenient but is by no means necessary. When knowing about common designs it is easy to generate test matrices that are appropriate for our needs. After collecting the data, regression analysis can be carried out in a program like Excel. If it is to be done on a regular basis it is probably more convenient to write scripts in MathWorks MATLAB®, R, or a similar environment. In MATLAB, regression models and statistics can be obtained using the functions REGSTATS or REGRESS and it is quite easy to write scripts that automatically generate residual plots, prediction plots, contour plots and so forth. MATLAB even has an optimization toolbox with a graphical user interface. For those who prefer dedicated software, Minitab®, JMP® and Umetrics' MODDE® are some common packages, each with its own advantages. In contrast to the others, JMP is available also for Mac OS X. These packages support more advanced designs than the ones introduced in this chapter. A notable example is the so-called D-optimal design, which can be used when experimental constraints make a completely orthogonal design impossible.

Let us finish the chapter with a summarized workflow for choosing designs and evaluating data from them. If the factors are categorical, two-level designs are appropriate. With continuous factors, such designs should at least be supplemented with center points to detect potential curvature. If a second-order model is to be fitted to the data, additional treatments are needed and central composite or Box–Behnken designs become suitable alternatives. Before actually collecting any data the orthogonality of the design should be investigated and confounding patterns established. After fitting a model to the data a number of diagnostic checks are made. The residuals should be normally distributed and have equal variance. This can be assessed using the residual plots introduced in Chapter 8. Non-normal residuals can often be amended by variable transformations. We should also confirm the absence of undesired time trends by plotting the replicates and the residuals versus time. After this the model is trimmed by removing unimportant terms while not letting R^2 decrease too much. This results in a more parsimonious model that shows the important relations in the data more clearly than the maximum model. If coded units are used the coefficients expose the relative importance of the factors. It is often useful to make a Pareto plot of the coefficients to obtain a graphical overview of the effect sizes. Finally, the model is used to plot, analyze and optimize the response.

9.10 Summary

- It is often necessary to design experiments with noise and background factors in mind to ensure that measured effects are due to the treatments of the experiment.

- One way to avoid background effects is to make *randomized controlled experiments*. These compare the experimental group to a control group, which is not manipulated. In medical studies the control group receives a placebo. As a further precaution such studies should be *blinded* or *double-blinded*.

- In a *crossover experiment*, the subjects receive a sequence of at least two treatments, one of which may be a control. After receiving one treatment they are crossed over to the other. Different groups receive the treatments in different order. The precision in the data increases as compared to randomized controlled experiments, as each individual is his or her own control.

- Several categorical factors can be tested simultaneously in *full factorial experiments*. They consist of orthogonal test matrices where all factors are varied simultaneously while still allowing complete separation of their effects. In particular, the two-level full factorial design tests all possible combinations of factor settings at two levels. These types of experiments also detect factor interactions.

- If factor interactions are present, the response to a change in one factor depends on the setting of another. This is a common situation, as many phenomena rely on appropriate combinations of factors.

- *Fractional factorial experiments* have reduced numbers of treatments compared to full factorial experiments. They are based on confounding one or several factors with high order interactions. A side effect is that other interactions or factors are confounded also. These designs are typically used for factor screening, since they provide less information than full factorial designs. The *resolution* describes the degree of confounding.

- *Plackett–Burman designs* constitute another class of screening design. The number of treatments is a multiple of four. When the number of factors increases, this makes it easier to find a test plan of suitable size, compared to fractional factorials. Since the main effects and two-way interactions are confounded in these designs they should be applied with caution.

- Continuous, numerical factors may produce curvature in the response. *Center points* can be added to two-level factorial designs to detect curvature but several additional treatments are required to fit a second-degree regression model to the data.

- *Response surface methods* combine design of experiments and regression modeling. Designs suited for such methods include *central composite designs* and *Box–Behnken designs*.

Further Reading

Box, Hunter and Hunter [4] provide one of the clearest and most practical guides to design of experiments. Montgomery [5] gives a more theoretical treatment in a popular book that is currently in its seventh edition.

Answers for Exercises

9.1–9.4 Solutions are given in connection with the exercises.
9.5 The number of runs doubles with each added factor, from four runs with two factors to 256 runs with eight factors.

9.6 *A-BCD*, *B-ACD*, *C-ABD*, *D-ABC*, *AB-CD*, *AC-BD*, and *AD-BC*.

9.7 The solution is given in connection with the exercise.

9.8 Main effects are confounded with two-way interactions.

9.9(b) Resolution V. Main effects are confounded with four-way interactions, and two-way with three-way interactions.

9.10 The solution is given in connection with the exercise.

9.11 $R^2 = 0.98$.

9.12–9.13 Solutions are given in connection with the exercises.

References

1. Fisher, R.A. (1951) *The Design of Experiments*, 6th edn, Oliver and Boyd, Edinburgh.
2. Box, G.E.P. (1990) Do Interactions Matter? *Quality Engineering*, **2**(3), 365–369.
3. Greenfield, T. (2009) Speech after receiving the George Box Medal. ENBIS 9, European Network for Business and Industrial Statistics, September 20–24, Gothenburg, Sweden.
4. Box, G.E.P., Hunter, J.S., and Hunter, W.G. (2005) *Statistics for Experimenters: Design, Innovation, and Discovery*, 2nd edn, John Wiley & Sons, Inc., Hoboken (NJ).
5. Montgomery, D.C. (2005) *Design and Analysis of Experiments*, 6th edn, John Wiley & Sons, Inc., Hoboken (NJ).

10   Phase I: Planning

To be playful and serious at the same time is possible, and it defines the ideal mental condition. Absence of dogmatism and prejudice, presence of intellectual curiosity and flexibility, are manifest in the free play of the mind upon a topic.

—John Dewey

This and the next two chapters present a methodology for research where many ideas discussed previously in this book are given a more concrete form. The research process is outlined using a number of activities and conceptual tools aimed at familiarizing the reader with ways of thinking that are useful in the various phases of research.

There is a potential danger in presenting research as a sequence of activities, since it could be interpreted as a fixed recipe for research. All research projects are unique and the path from question to answer tends to vary greatly. Still, most research tasks go through similar phases and the aim of the methodology is to help the reader reflect on the various activities that take place between these two points. This is not a cookbook. It is a series of examples and exercises meant to expand and elaborate the ways in which you think about your research.

This chapter focuses on the phase that takes place before any data are collected. In many ways, most of the intellectual effort of a research project goes into this phase; tools for generating and evaluating hypotheses will be of central interest.

10.1   The Three Phases of Research

Although individual research projects differ, they typically move through three common phases. In the first phase, a research question is formulated and a strategy is laid out for answering it. The research question is often connected to one or several hypotheses. We will call it the planning phase. Once a promising approach has been developed it is time to gather evidence to either support or refute the hypothesis. This is the data collection phase, aiming to obtain data of sufficient quality to draw robust conclusions. It involves investigating the capability of one's measurement system, in some cases even the actual development of a measurement system, as well as identifying suitable procedures for measurement. In the final stage the data are analyzed and connected back to the research question. The findings are also related to the current knowledge to define a scientific contribution. This third phase will be called analysis and synthesis. Each of the final three chapters of this book is devoted to one of these three research phases, which are schematically represented by Figure 10.1. The process is depicted as circular or periodical, since the new

Experiment!: Planning, Implementing and Interpreting, First Edition. Öivind Andersson.
© 2012 John Wiley & Sons, Ltd. Published 2012 by John Wiley & Sons, Ltd.

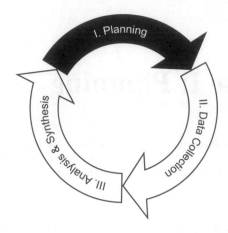

Figure 10.1
The three phases of research. Research is cyclic, since results from one research task often give rise to a new one.

knowledge that results from one research activity often gives rise to new questions and new research ideas.

This chapter introduces a number of tools and techniques for generating and evaluating hypotheses. Many of these tools were originally intended for solving quality problems, for example in product development organizations or production plants. In many respects, quality work is similar to research. Imagine a team of engineers that is assigned to solve a quality issue. It could be anything from a product not meeting the specifications to sudden malfunctioning, like mobile phones switching themselves off or excessive oil consumption in a car model – anything resulting in customer complaints. The team's task is to find and fix the cause of the problem as soon as possible. This can be very demanding as there are often many potential causes for such problems. They often have to consider everything from the production process to the environment where the product is used. By thinking of as many potential explanations as possible and excluding the possibilities one by one, they gradually converge on the real cause. This is very similar to what researchers do. Faced with a phenomenon that is not understood, they attempt to find potential explanations for it, and by investigating various hypotheses they approach the true scenario.

We will borrow a number of techniques from the quality toolbox. They are very simple and at first sight it might seem naïve to use them to teach scientific method. Research projects are, after all, often complex and intellectually very demanding. It should be stressed that the tools are not very interesting in themselves. They are to be seen mainly as pedagogical tools. What is interesting is the intellectual process that starts when you use them. I do not mean to imply that working scientists use these tools specifically but applying them will familiarize the reader with *ways of thinking* that are useful in research.

It is also important to point out that application is essential for all learning. If you explain the experience of scuba diving to a person who has grown up in a desert and has never seen the ocean, you may only convey a vague picture. To really understand what diving is like, that person must go to the ocean and experience it at first hand. The exercises in these chapters work exactly like this; they will not affect your thinking about the research process until you have experienced how they apply to problems that are relevant to you.

To make the tools less abstract I am using two real experiments as examples. One example is from biology and one from engineering, to show that the thinking applies to different fields. It should be noted that the tools were not actually applied in these experiments. They are only used as vehicles for explaining how the tools may be applied to real research problems. You may or may not use these tools in the future but when you have tested them a couple of times you should have obtained a more intuitive understanding of some important aspects of the scientific process.

Before we discuss the techniques, let us take a look at the experiments that we are going to follow over the last few chapters.

10.2 Experiment 1: Visual Orientation in a Beetle

Scarabaeus zambesianus, an African dung beetle, flies off at sunset to forage for fresh dung. Once it finds some, it quickly forms a ball of dung with its front legs and head and rolls it away in a straight line – the most efficient path for escaping aggressive competitors for food in the dung pile. Apart from its peculiar eating habits, it also has an interesting ability to orient itself using the pattern of polarized light that forms in the sky at sunset. As shown in Figure 10.2, a dung beetle pushes the balls backwards with its hind legs. Despite being unable to see in the direction that it is moving, it maintains its departure bearing over a long distance. The beetle does this by using specialized photoreceptors in its compound eyes that sense the direction of polarization in the sky.

The reason that the sky has a polarization pattern is that sunlight is scattered off molecules in the air. This process is known as Rayleigh scattering and is also the reason why the daytime sky is not pitch black but has a soft blue glow. For the reasons explained in Box 10.1, the degree of polarization varies across the sky in relation to the direction of the sun. A band of highly polarized light stretches across the zenith at sunset, with the degree of polarization decreasing gradually in the directions towards and away from the setting sun. The location of the band is shown schematically in Figure 10.4, where the dashed semicircle represents the dominant direction of polarization. Though invisible to us, this pattern gives the dung beetle a very palpable sense of direction.

Figure 10.2
Dung beetle pushing a ball away from the dung pile using its hind legs. © Marie Dacke.

Box 10.1 Polarized skylight

- Light can be seen as an electromagnetic wave. The *polarization* describes how the electric field is oriented in the wave.
- The dashed line in Figure 10.3 represents a sunray. It has one horizontal and one vertical polarization component, as indicated by the crossed arrows on the line. Since there is no dominant direction of polarization, this light is defined as *unpolarized*.

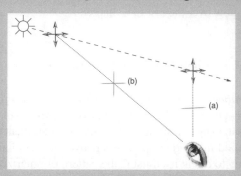

Figure 10.3
The dashed line represents a sunray. Black arrows represent the polarization components of the sunlight. Grey lines represent the polarization components of the scattered light. Sunlight scattered from the direction of the sun contains all directions of polarization and is thereby unpolarized. Light scattered at right angles to the direction of the sun is polarized, as only one polarization component can be scattered at this angle.

- The polarization is always perpendicular to the direction in which the wave propagates. It cannot be parallel to this direction.
- For this reason, light scattered at right angles from the sunray (a) only contains one polarization component. The vertical polarization component cannot be scattered in the vertical direction. Since (a) has a dominant direction of polarization, this light is defined as *polarized*.
- Light scattered along the direction of the sunray (b) contains both polarization components. Both components are scattered because none of them is parallel to the direction of the wave. This light is therefore unpolarized, just as the sunlight.

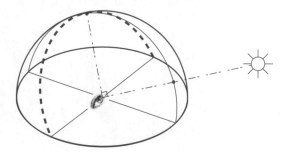

Figure 10.4
The sky location of the band of highly polarized light (dashed line) in relation to the sun.

By monitoring the beetles' movements under the nighttime sky Marie Dacke and coworkers found that, on moonlit nights, the beetles rolled their balls radially in straight lines from the dung, but on nights without a moon they entered on a more or less random walk about the pile. This made them wonder if the beetles were able to use also the polarization of scattered moonlight for orientation. This light being about a million times fainter than that from the sunlit sky, it was by no means obvious that it could be used for orientation. To confirm that the beetles used the polarization and not the moon itself, they hid the moon from view and put a polarizing filter over a dung-rolling beetle. Such a filter transmits one polarization component of the skylight and blocks the other. When the filter's direction of polarization was perpendicular to that of the sky's polarization, the beetles turned close to 90 degrees when entering under the filter. If the filter was oriented in parallel with the sky's polarization, the beetles maintained their bearing under the filter [1]. This was the first demonstration of an animal using the polarization pattern in the moonlit sky as a nocturnal compass. This ability is obviously advantageous as it allows the beetle to extend its foraging time well after sunset. The method is useful also under partly cloudy conditions when the moon itself may be hidden behind a cloud.

This was clearly an important finding, so it was with mixed emotions that Dacke discovered on a later occasion that, suddenly, the beetles could orient themselves also on nights without a moon. How was this possible? She started toying with various hypotheses. It was already clear that the beetles could use several cues for orientation, such as the sun, the moon, and polarization patterns in the sky. If one cue were missing, it would use another. In the laboratory she had shown that the beetles could orient to light bulbs. If one bulb was switched off and another switched on, the beetle would change direction. Leaving the bulb on the beetle would proceed in a curved path around the bulb, probably in an attempt to keep the light source centered in a region of its compound eye. As the moon and sun are infinitely far away this method would result in a straight path in the field.

Was it possible that they used the stars for orientation? Larger animals had been shown to do so, but since a small compound eye can only collect a comparably small amount of light, most stars would not be visible to the beetle. On the other hand, seeing only a few of the brightest stars would be sufficient for orientation. The Milky Way was another possibility. It had been low in the sky during her first experiment but was almost at the zenith now. But to test whether the beetles used bright stars or the Milky Way for orientation she would have to switch parts of the night sky on and off, which of course would be impossible. Or would it? At this point, Dacke decided to call the planetarium in Johannesburg with a singularly odd request.

She wanted to bring her beetles – and fresh dung – into the planetarium to see how they reacted to different cues in the night sky that, in a planetarium, could be projected onto the domed screen at will. This would make it possible to investigate how the beetles reacted to various subsets of the brightest stars or to the Milky Way. Rather unexpectedly, the management of the planetarium agreed that this was an excellent way to use their facility.

The research team would also set up a specially designed, two-meter wide wooden arena in the dome. They did not want to monitor the beetles using a camera, as the legs of the camera stand could work as landmarks for orientation. The arena was designed so that nothing but the planetarium dome would be visible to the beetles, which were placed at its center through a trap door from underneath. A number of collection troughs were mounted around the perimeter of the arena for receiving the ball when it was pushed over the edge.

The straightness of the path was measured by monitoring the time between placing a beetle at the center and the audible bump of the ball falling into the trough. Previous results had shown that a random walk could take more than five minutes while a straight path to the edge would only take about 20 seconds.

The planetarium experiment resulted in another unique discovery. It showed that *S. zambesianus* could, in fact, orient to the Milky Way, adding another method of orientation to its already impressive repertoire. This was the first convincing demonstration of using of the Milky Way for orientation in the animal kingdom, as well as the first evidence of star orientation in any insect [2].

10.3   Experiment 2: Lift-Off Length in a Diesel Engine

In a diesel engine, the flame is always found some distance from the fuel injection nozzle. This distance is called the lift-off length and, as explained in Example 5.1, it is useful for understanding trends in soot formation in such engines. The lift-off length is generally assumed to stabilize shortly after auto-ignition and remain stationary until the end of the fuel injection period.

Most investigations of how the lift-off length is affected by various parameters are conducted in spray chambers with optical access. These are typically large chambers where important conditions, like pressure and temperature, can be precisely controlled over wide ranges. This is obviously a great advantage from a data collection point of view, but a potential drawback is that spray chambers are not very similar to the combustion chambers of real diesel engines. Due to the limited volume in an engine's cylinder, the combustion inevitably increases both the temperature and pressure, but they remain relatively unaffected in a large spray chamber. Furthermore, to simplify measurements, spray chamber experiments are usually made with single-hole fuel injectors, while injectors in engines often have six to eight nozzle holes. Since spray chamber experiments ultimately are made to understand the processes in real engines, it is important to understand how these differences affect the lift-off length.

In contrast to spray chambers, optical engines maintain most of the features of real engines, making it possible to investigate the effects of the differences. In these engines, parts of the combustion chamber have been replaced with transparent parts. A drawback is that the optical parts put an upper limit to the in-cylinder pressures; this limit is often lower than those allowable in spray chambers.

Clément Chartier had found that lift-off lengths were often shorter in optical engines than in spray chambers. When experimenting with an eight-hole nozzle in an optical engine, he also discovered that the lift-off length did not stabilize during injection – it was constantly drifting towards the nozzle [3,4]. To understand the reason he used an empirical formula describing how the lift-off length is affected by various parameters, which was based on an extensive database of measurements with single-hole nozzles in a spray chamber. The formula clearly showed that the temperature of the air surrounding the fuel spray was a dominant factor.

Firstly, he calculated two interesting temperatures in the combustion chamber of his optical engine. One was the temperature of the fresh air that had not participated in combustion, the other was the mean temperature in the combustion chamber – an average including the temperatures in the flame, the burned gases and the fresh air. Then he used the empirical expression from the spray chamber "backwards", to calculate the temperature required by this formula to yield the lift-off length he had measured. This temperature was substantially higher than the other two, indicating that the air surrounding the jet in the engine contained large amounts of hot, burned gases. An idea for an experiment started to form.

When injecting a single spray into a very large volume, mostly air is mixed into the spray close to the nozzle. For this reason the burned gases do not affect the lift-off length. In the engine, however, where several sprays are cramped into a much tighter space, it seemed like a substantial portion of the gas around each spray consisted of burned gases. This raised the question if the spacing between the sprays affected the lift-off length. Chartier decided to find out.

The hypothesis was that two burning sprays would constantly replenish the space between them with hot combustion products. This would gradually raise the temperature and push the lift-off closer to the nozzle. The smaller the inter-spray spacing, Chartier argued, the more the temperature would rise. To test this idea, he compared two different injector nozzles in the optical engine. One had four holes in a symmetric configuration, producing sprays that were at right angles to each other. The other also had four holes, but in an asymmetric configuration. One of the sprays was wedged in at 45-degree angles to its closest neighbors, whereas the opposite spray enjoyed a spacious 135 degrees to its companions. Figure 10.5 shows chemiluminescence images of the burning sprays taken with the two nozzle configurations in the engine. Keeping the number of holes and the total fuel flow rate constant in the two configurations allowed a direct comparison of the effect of the inter-spray spacing. This is important since changing the number of holes would have changed the amount of fuel injected per unit time, and thereby altered the temperature history in the cylinder.

In addition to the inter-spray spacing, Chartier's experiment included other parameters that affect the lift-off length. It turned out that, in the optical engine, the effect of the air temperature was much weaker than spray chamber results indicated, and the inter-spray spacing actually had a stronger effect. Chartier attributed both these observations to the

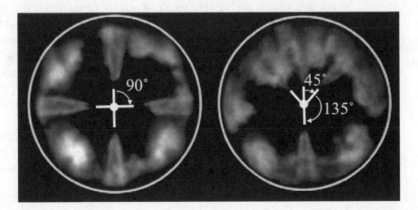

Figure 10.5
Chemiluminescence images of burning fuel sprays in an optical diesel engine. Imaging takes place from the bottom, through the transparent piston. The nozzle is centrally positioned and the white circle indicates the boundary of the combustion chamber. Left: symmetric nozzle. Right: asymmetric nozzle. © Clément Chartier.

presence of hot, burned gases between the jets. It was also clear that the conditions varied between the combustion cycles in the engine, which resulted in substantial variability in the lift-off data. Such cycle-to-cycle variation is a well-known phenomenon in engines that is not accounted for in spray chambers, which are designed to produce extremely stable conditions.

Spray chambers are very valuable tools for understanding fundamental properties of sprays, but the results showed that one must proceed with caution when applying results from spray chambers to the analysis of real engines.

10.4   Finding Out What is Not Known

The first step of planning is finding out how you can contribute new knowledge to your field. To do this, you must determine the boundaries of the current knowledge. As research papers will not be published if they do not contain new information, it is important to find out how to make a novel contribution. This may seem challenging to new Ph.D. students, since their whole field of research is new to them. As a rough rule of thumb, the first half of your time as a Ph.D. student will be used to reach the research frontier in your area. During the second half you are expected to make "an original contribution to knowledge", which is what the Ph.D. is awarded for.

Getting familiar with the entire bulk of knowledge accumulated in your field over the years is a formidable or, quite likely, impossible task. You need to be selective. It is important that academic supervisors direct their students to research papers that have been influential. Get the "classics", read them and reflect on them. You should regularly browse the scientific literature in your field to keep up to date with the latest developments. It is also important to be in touch with your part of the scientific community. The daily discussions with supervisors, other faculty members and more experienced Ph.D. students at your department are important, but it is just as important to listen to their discussions with other researchers. Conversations on a research topic will often give you valuable clues to "what is hot and what is not". Every visit to a research conference is a golden opportunity to get an overview of research problems that are currently addressed and how they are studied. By taking in, processing and working with all this information, you gradually build a theoretical framework for your own research – a mental picture to orient yourself by. This picture will make it easier for you to see where and how you can make a contribution. Although the demand for novelty can be distressing for new Ph.D. students, their confidence tends to increase with time. As your mental picture gets clearer, you realize that making an original contribution is not an insurmountable obstacle.

Apart from knowing the boundaries of the current knowledge, it is also helpful to know something about what constitutes novelty in a piece of research. A list of the components of research is given below, followed by a discussion about the ways in which they can be new. Research results can be said to be the output from:

- Data
- Methods and techniques
- Research questions
- Research areas
- Analyses and syntheses.

Obtaining good and reliable *data* is central to all research. A unique experiment will provide unique data, but this is not sufficient to fulfill the need for novelty. New data that support an established theory, like Galileo's law of free fall, are not an especially valuable scientific contribution. Though such data are good news for the theory, confirmation of a thoroughly tested theory does not add much to our knowledge – unless it is confirmed under conditions where it is not expected to be valid. Data that show discrepancies with current theory, on the other hand, are interesting, since they may point to limitations in the current knowledge. But not all experiments are made to test theories. Their purpose may be to characterize a system or to investigate some aspect of its workings. The crucial point is that your data should bring new, useful information to your research community.

New *methods and techniques* can make it possible to study new aspects of a problem. They could be novel measurement techniques or new methods for analyzing data. Applying a known technique to a new area can be a sufficient criterion for novelty, as well as applying a new technique to an old area. The method does not even have to be applied since, in some fields, the development of new methods and techniques is the very subject of research. It is also worth mentioning models under this heading. Models are mathematical representations of systems, such as the earth's atmosphere, the combustion chamber of a diesel engine, an evolutionary process, or the traffic situation at a highway junction. Models are often complex, containing a large number of functions to account for various aspects of the system's properties. They are often based on a variety of simplifying assumptions. In contrast to theories models are not generally valid, but may be useful under well-defined conditions where they have been tested and calibrated. In most cases, models are used in computer simulations of a system's behavior, making it possible to study phenomena that are impossible or unpractical to study experimentally. You may, of course, make a novel contribution to research by developing a new model or carrying out a new modeling study.

Research questions are the very starting points of research. Finding a question, small or big, that has not yet been answered within your research area is one of the most promising ways of expanding the current knowledge. This is echoed in the quote at the beginning of Chapter 2, saying that it is less important for a researcher to know the right answers than to find out what the right questions might be. Finding new research questions may sound difficult but is often easier than one thinks. Looking at how research is done in your field, you will probably find that people tend to approach certain aspects of certain types of problems using certain methods. When you have identified such common characteristics, change your perspective and try to see what is *not* there. With a bit of creativity, you will probably find numerous questions that are short of attention.

Discoveries sometimes lead to new *research areas*. To name just one example, one of the central problems in physics at the turn of the last century was that there was no coherent theoretical treatment of heat radiation. Max Planck eventually discovered that this problem could be solved by treating light as discrete packages of energy; this was the starting point of quantum theory [5]. Very few can hope to make such grand contributions but, luckily, new research areas often occur on a much more modest scale. They can be the result of advances *within* an existing discipline. For example, a new method could make new types of studies possible. An interesting new research question or a new synthesis of results could also displace the interest towards an area that has not previously been studied. New research areas also emerge *between* existing disciplines. Recent decades have seen a boom of interdisciplinary research areas, combining methods and problems from several subjects.

Finally, novelty can consist of a new *analysis* or *synthesis* of existing evidence. As discussed in Chapter 5, questioning the validity of evidence and conclusions is an important way in which knowledge develops. Research papers are brief and often focus on a very specific question. Bringing together and comparing results from several previous studies that bear on a more general question can sometimes bring forth interesting new aspects of a problem, highlight limitations in a common approach or even point to misconceptions about a certain problem.

There are obviously many ways to be original without revolutionizing an entire discipline. Novelty is an imperative demand in all research but you will find that it is usually not a reason for worry. You could be original in testing someone else's idea or even by continuing a previously original work. In fact, previous work is one of the most common inspirations for new research tasks. Your own results will have a tendency to result in new research ideas, simply because your analysis involves a great deal of reflection. When you pursue these ideas, you are successively expanding the knowledge within your field.

> **Exercise 10.1:** Use the list above to determine the ways in which the two example experiments make novel contributions to science.

Let us look at some methods for generating research ideas. The overall goal of science as it is defined in this book is to explain the world. It is therefore useful to make a habit of challenging your ability to explain phenomena within your field. One way to do this is to repeatedly ask yourself the question, *"Why is that so?"*, when you encounter statements about a phenomenon in a research paper or during a discussion. Anyone who has known a four-year old knows how frustrating such questions can be, because you very quickly tend to run out of answers. For example, if the question is, "Why is the sky blue?", the answer is that sunlight is scattered from molecules in the air. "But why is it *blue*?" Well, it is because this process is more efficient at shorter wavelengths, so blue light is scattered more than red. "But *why* is that so?" At this point you need an explanatory theory to provide an answer, which has to do with how photons interact with molecules. Most parents probably resort at a much earlier stage to saying "because that's the way things are", which, to a scientist, has a rather Aristotelian ring to it. The point of this technique is that the answers tend to move through a succession of descriptive statements before reaching a level where our *understanding* of a problem is finally challenged. Upon reaching that level you will know whether you, or anyone else come to that, can explain the phenomenon. That is why this technique is useful for finding out what is not known and, thereby, for generating new research ideas. For lack of a better term, we could call this technique *why-analysis*. When writing a scientific paper the same technique can be used as a quality control: act as the curious four-year old with every statement in your conclusions to see if you really can *explain* what you claim.

> **Exercise 10.2:** Apply why-analysis to a phenomenon that is relevant to your research to see if you can find aspects that are not completely understood.

When you have explored the boundaries of the current knowledge and found a promising problem to work with, it is time to start generating hypotheses. In the hypothetico-deductive approach to research these tentative explanations of the observed phenomenon are the foundations of experiments, conducted to either support or refute the hypotheses. As described in Chapter 6, experiments may or may not be based on hypotheses. Non-hypothesis-based experiments are sufficient if you are only interested in connecting a cause with an effect. The hypothetico-deductive approach is favored here, since hypotheses allow us to draw more complex and interesting conclusions from our data. Depending on the nature of the hypotheses they may even be used to develop explanatory theory. Before looking at tools for generating hypotheses, however, we will discuss the importance of limiting our ambitions.

10.5 Determining the Scope

Most problems are complex. One of the most important tasks in planning is, therefore, to break problems down into smaller pieces that can be handled. A Ph.D. project that aims to solve world hunger, find a general cure for cancer or to find an infinite source of clean energy will almost certainly fail. This is not because these problems are impossible to solve but because they are too extensive for a limited project. The problem is very old. How does one eat a mammoth? It obviously has to be done one bite at a time. (In the end, it turns out, we actually managed to eat them all.)

The scope defines the limits of the work to be done. It involves choosing what to do and, also, what not to do. The simple reason for keeping ambitions at a realistic level is that it increases the chances of success. The problem of a Ph.D. thesis should not be too difficult to solve. The purpose of research studies is to learn the craftsmanship of research. Ph.D. students demonstrate this knowledge by making an original contribution to knowledge, but if they were expected to revolutionize their research fields, very few doctors would be graduated. This is not to say that the research problem should be trivial. The point is that you learn better by doing something that can actually be done. Almost every important problem must be broken down, scoped and treated bit by bit in order to reach a solution.

Scoping a research project has three important aspects. The first is the scientific one of identifying the most interesting parts of the problem. This defines the *goal* of the project. The other two aspects are practical ones, since all work is necessarily conducted with limited *resources* in a limited *time*. Project managers often illustrate these three limitations by a triangle where one side cannot be changed without affecting the others. If the goal becomes more ambitious the project will require either more time or more resources. The latter are often fixed, since research projects have a definite level of funding for a fixed period. What remains is to decide how to spend them. Should time and money be devoted to developing new methods or building new equipment? Do you need comprehensive information in a limited area or is it valuable to cover more aspects of the problem in less detail? Break the problem down to individual tasks, focus on them one at a time, and choose a track that is both feasible and aligned with your goal. Once the track is settled you have locked the triangle. When new ideas pop up, as they tend to do in research projects, you must reprioritize. You cannot add activities without taking others out, unless more time or money is granted.

Incidentally, you will have realized by now that the word "problem" is not a negative word in research. The research problem is the challenge that sparks the scientist's imagination – the very motive for doing research. When you have identified interesting problems that can actually be handled, it is time to formulate the individual research tasks. The process starts by generating hypotheses.

10.6   Tools for Generating Hypotheses

The first step in generating hypotheses is to trace an observed phenomenon back to potential causes. This requires an active imagination. The theoretical framework that you have developed by working in your field will affect how you see and approach research problems. As discussed in the section about creativity in Chapter 5, this tends to make you search for certain types of solutions that give you a sense of familiarity and security. In other words, your expertise tends to close you off from alternative lines of thought. Needless to say, this can be an obstacle when looking for new ideas.

It may seem contradictory to first stress the need for developing a mental map of the research problems in your field and then explain that this map is an obstacle for your thinking. The truth is that creative work is contradictory. To be able to do anything of value you must develop the map and use it, but it is also important to be able to free yourself of this intellectual luggage in situations where fresh thinking is required. Remember that the map is not fixed. It is a product of your own thinking, and thereby varies from the maps of other researchers. It is a dynamic map that changes and grows as you learn from experience.

It is often useful to visualize a problem in different ways. One way to do this is the so-called cause-and-effect diagram, introduced by Kaoru Ishikawa as a quality control tool in the 1960s. Due to its shape it is sometimes called a fishbone diagram. Here, we will use the term *Ishikawa diagram* to avoid confusion with cause-and-effect tables, which will be introduced later. The diagram allows the cause-and-effect relationships in a process to be summarized and it is, therefore, useful also in scientific research. When constructing the diagram, we are forced to think of as many different explanations for a phenomenon as possible, which keeps us from settling with the first and most obvious idea.

To make a cause-and-effect diagram, take a large piece of paper and write the effect or phenomenon that interests you in a box at the far right of the sheet. Then draw an arrow across the paper from the left to the box on the right. If you are studying a process this axis represents the timeline, but the diagram can also be used to analyze phenomena that does not have an obvious temporal dimension. Now, think carefully about all the possible main causes for the effect. When applying the diagram to an engineering process these are often the "5Ms"; man, machine, material, method, and measurement. In other situations you may choose other broad headings that apply to your particular problem. In Experiment 1, for example, the beetles rolling straight could potentially be explained by a wide range of sensory perceptions, like smelling, hearing and so on. Enter such main causes into the diagram with smaller arrows pointing onto the main axis. If the diagram describes a process you should start from the left and follow the order of the process. The result is a diagram that roughly resembles a fishbone or a Christmas tree that has fallen over to the right (Figure 10.6). It shows one possible Ishikawa diagram for Experiment 1. It is useful to try various methods of classification, for example process order (time), physical location (space), or any other classification that comes to mind. In this case, various visual cues were chosen as headings. The next step is to look at each of the branches in turn and try to find as many subcauses as possible. Enter them into the diagram with arrows pointing onto the branch. Remember that Figure 10.6 is just one conceivable diagram for Experiment 1. Another person could come up with other ideas or prefer different classifications. There

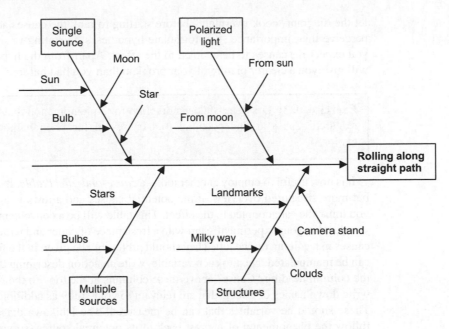

Figure 10.6
Ishikawa diagram of potential causes explaining why the dung beetle can roll a ball along a straight path.

is no right diagram; the mental process that is started when drawing the diagram is more important than the diagram itself. One important goal is to move beyond the obvious ideas suggested by your current knowledge and explore new perspectives on the problem.

As Ishikawa himself points out, there are no specific rules for drawing the diagram. The important thing is to break the causes down using sub-branches, sub-sub-branches and so on, to the point where you have identified potential causes that can be *investigated*. Why-analysis is a very useful tool in this process [6].

Sometimes, the research problem focuses on a process, as in Experiment 2. Here, the effect is the result of a sequence of events experienced by the fuel, from entering through the nozzle to ending up as combustion products in the exhaust port. Other research problems focus on phenomena that have no obvious timeline. Experiment 1 is a good example of this, as the beetle's orientation could be explained by a number of visual (and potentially other) cues, but it would not make any sense to order them on a time axis. The diagram is still a useful way to structure your thoughts but you may prefer to put the effect at the center of the paper and organize the causes like spokes and subspokes around it, as in a mind map. That is really what the diagram is – a tool for organizing and visualizing your thoughts. Ishikawa preferred the timeline structure with angled arrows because it gives a visual impression of following a river back to its source [6].

It could be useful to interact with other people when constructing the diagram, or even to make it a team exercise. This will often increase the number of ideas and perspectives considered. If you run out of paper space you could transfer a branch of the diagram to a separate sheet and elaborate it further. When you are satisfied, put the diagram away and let your subconscious process the information. It is important to abstain from negative comments during the brainstorming phase, as it tends to kill the creativity. It is better to let the diagram "cook overnight" before starting to rank the causes according to how you perceive their importance. You formulate hypotheses by stating in simple sentences how you expect the causes to be coupled to the effect. Apart from the hypotheses, the diagram will give you a clearer picture of your problem than you had before.

> **Exercise 10.3:** Draw an Ishikawa diagram for a phenomenon that is relevant to your research. Can you find potential root causes of the phenomenon that are not yet explored?

It is now useful to employ another tool – a *cause-and-effect table*. It gives a less graphical but more structured overview of the potential causes and allows you to elaborate on how you think they are coupled to the effect. This table will be a convenient starting point when you develop your experiment. Start with a fresh piece of paper and write the most promising causes in a column on the left. They should preferably be given in the form of variables that can be manipulated. Next to each variable, write an action describing the manipulation. Put the column heading "Cause" above these columns at the top. To the right of this heading, write down names of effects that are relevant to your study as additional column headings. These should be variables that can be measured. The Ishikawa diagram allowed you to follow the phenomenon of interest back to its potential source, so you could in principle use several intermediate branches between the phenomenon and its final cause as headings. In that case it is convenient to write them in an order that describes a causal chain from the causes on the left to the phenomenon on the right. You could put down any other effects that you expect the causes to have as well, as these may provide additional ideas for your experiment. It is important to stress that research problems are different and, accordingly, the structure of the diagram may differ.

After listing potential causes and effects in the table, it is possible to make simple thought experiments. Look at the causes on the left, which are given as variables and actions. Try to imagine what these actions do to each of the effect variables on the right. Does the effect decrease when the cause variable increases? If so, write a "−" sign in the effect column. If the effect and cause are positively correlated, write a "+" sign. If you expect the effect to be very strong, you could put a double "+" sign or, if you think the effect is weak, put the sign within parentheses. Not all causes will affect all effects, so it may be appropriate to put a zero in a column. If it is difficult to anticipate an effect, you could use a "?", and you can use other symbols to represent a quadratic dependence and so forth. If the columns of the table describe a causal chain between the root cause on the left and the final effect on the right, you may have to consider all the columns to the left of a given column to estimate what that particular effect will be.

Table 10.1 shows a possible cause-and-effect table for Experiment 1, which was developed from parts of the Ishikawa diagram in Figure 10.6. It contains clues to how various hypotheses can be tested. For example, if we assume that the beetles use the polarization pattern from scattered moonlight for orientation, they would not be able to roll straight on nights without a moon. This is represented by a "+" sign in the "random walk" column, as that effect is expected to be correlated with the absence of moonlight. The sky polarization can be used for orientation if the moon is up, even if it is hidden from view. This is indicated with a "+" in the "constant direction" column. The final cause under the polarized light heading is the use of a polarization filter, which could affect the beetle's sense of direction and make it change direction. As described in Experiment 1, all of these observations were

Table 10.1 Cause-and-effect table based on parts of the Ishikawa diagram in Figure 10.6.

CAUSE		EFFECT		
Variable	*Action*	*Random walk*	*Constant direction*	*New direction*
Single source	Hide it	+	−	0
	Move it	0	−	+
Polarized light	Moonless night	+	−	0
(from moon)	Hide moon	0	+	0
	Polarizing filter	0	−	+
Milky Way	Remove it	+	−	0
	Rotate it	0	−	+

made in the field and thereby support the hypothesis that the beetles use the polarization properties of scattered moonlight for orientation.

After filling out the whole table you have a qualitative description of how you anticipate the root causes to be coupled to the phenomenon of interest. In some cases the relationship is expected to be strong and in other cases the outcome is less predictable. The causes that are strongly coupled to the effect are generally the most promising ones to test experimentally, as a strong effect will provide a more unambiguous test of your hypothesis than a weak one. If an effect that was expected to be strong turns out to be absent in your experiment, you may discard your hypothesis and move on to another one.

Again, this is just one way of constructing a cause-and-effect table. You do not have to adhere strictly to this format. Every research problem is unique and yours may invite other ways of organizing causes and effects. Remember that the techniques presented in this chapter are tools to aid *your* thoughts. They should be adapted to suit the problem, not mindlessly repeated.

> **Exercise 10.4:** Make a cause-and-effect table based on your Ishikawa diagram from Exercise 10.3. Do you get any ideas for interesting experiments?

Our brains seem to prefer concrete, visual scenarios to abstract reasoning. If you are studying a process, one method for visualizing the research problem is to close your eyes and imagine taking part in the process. In Experiment 2, for example, the fuel goes through a number of events in the cylinder. If we imagine being a fuel molecule and try to picture how it experiences all these steps, we will be forced to consider the parts of the process in detail. Our mental walk through the process can be documented in a *process diagram*. This is basically a set of boxes that represent the steps of a process, connected by lines or arrows. There are standards for how these boxes should be drawn to describe different types of events but for your private use you may draw the boxes whichever way you find useful. Figure 10.7 shows a simple process diagram describing the fuel history in Experiment 2. The research question in this experiment was whether significant amounts of hot combustion products were mixed into the spray or not. The diagram could be made considerably more complex but this version is sufficient to convey a general idea. When there are alternative results of a specific process step, this is represented by diamond-shaped boxes with conditions written into them. Depending on whether a condition is fulfilled or not, the process may follow one of two paths, represented by lines labeled with the words "yes" and "no". Visualizing the process like this can highlight critical events and suggest

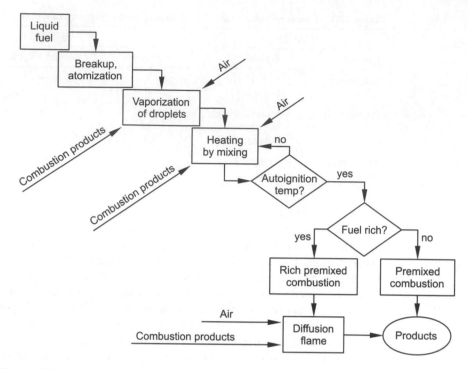

Figure 10.7
Simplified process diagram of the fuel history from entering the combustion chamber through the nozzle to ending up as combustion products in the exhaust port.

ideas for experiments. If we, for instance, wish to confirm that hot combustion products are present and play a role in the auto-ignition step, Figure 10.7 tells us that these hot products will affect the vaporization step too. This gives us an additional method of supporting the hypothesis that combustion products are mixed into the spray. Without the diagram this possibility might have been overlooked.

Exercise 10.5: Draw a process diagram for a process relevant to your research. If you do not work with processes, choose one to explore. It could, for instance, be the history of a tomato stain on a shirt in a washing machine, the formation of a coprolite (fossilized dinosaur dung), or something completely different.

The tools introduced in this section are intentionally quite simple and intuitive, since the focus should be on the research problem and not on the tools. Their purpose is to help us consider more possibilities than we would if we settled with our first idea. Allow plenty of time for working with them. Even though the techniques are simple, exploring the various perspectives is hard work if it is done seriously. When using these tools, the voyage is more important than the destination. You could develop your own thinking techniques too. Any method that works with your way of thinking is of course useful.

10.7   Thought Experiments

Once hypotheses have been generated it is time to start to evaluate and test them. The final test is often an actual experiment but many ideas can be evaluated and refuted just by thinking about them. Thought experiments are means for exploring the logical structure of a problem and can be very powerful tools for testing ideas. They use imaginary scenarios to understand features of the real world.

We often think that we are unable to make statements about a certain problem if we do not have basic knowledge of the specific problem area. For example, you may not think that you are able to find out what is wrong in the central heating system in your house if you are not a plumber. As a matter of fact we all have general knowledge about the world, both from experience and theoretical knowledge. Based on this we can exclude possibilities, since we also know how the world does *not* behave. You intuitively understand that pipes transporting water to the radiators of your heating system should be warm, for instance, and that pipes leading away from them should be slightly less warm. Such knowledge can be used for troubleshooting even without detailed knowledge of the internal workings of heat pumps, furnaces or what have you to heat your house. In the same manner, your general knowledge can be used for troubleshooting theoretical ideas. If an idea inevitably leads to a paradox, there is strong reason to believe that the idea is wrong. The cause-and-effect table is a very simple example of trying to foresee the consequences of ideas by reasoning. We have also seen more complex examples of elegant thought experiments in this book.

In Chapter 4 we saw how Galileo refuted Aristotle's idea that heavy objects fall faster than light ones in free fall using nothing but logical arguments. Say that a light object falls at a speed of four units, he said, and that a heavy object falls at a speed of eight. By attaching the two objects to each other, the heavy object will be retarded by the light one, and the light object will be sped up by the heavier. This means that the two objects together, though heavier than the parts, will fall at a speed less than eight. In other words, Aristotle's claim is not logically consistent. Or, as a sports commentator might put it: "Galileo–Aristotle, 1–0".

Another thought experiment is described in Example 6.1, where William Harvey tested Galen's claim that blood is continuously formed from food in the liver and consumed in the limbs and organs of the body. Even a conservative estimate of the rate at which the heart pumps blood into the body would amount to far greater quantities than could possibly be provided by eating in a day. He also noted that a butcher could empty an animal of blood in less than fifteen minutes, which was logically inconsistent with the antique idea.

Thought experiments often employ specific scenarios or perspectives to test an idea. They can involve simple quantitative reasoning, as in the examples above, to see if a general statement applies to a specific situation. If the idea can be expressed in mathematical form it can also be tested using order-of-magnitude estimates, where powers of ten are put in the place of important variables to see if the resulting magnitude is consistent with the claim. This sort of estimation is often called a *back-of-an-envelope calculation*, a term that goes back to the physicist Enrico Fermi who was known for his ability to make good approximations using little or no data. They often involve *guesstimates*, a term coined by statisticians to describe that a quantity is very roughly approximated or even guessed for lack of accurate information. It is of course important to state the assumptions and approximations behind all estimates. Another common method is to use extremes: letting variables in a mathematical expression approach zero or infinity to see if paradoxes occur.

The basic idea is that general claims must apply to specific situations. If they do not, they are not generally valid and may be discarded.

Let us look at how back-of-an-envelope calculations can be used in thought experiments. Most of us were able to refute the Santa Claus hypothesis at quite an early age. It claims that an outsized elf visits some hundred million homes in the western world on December 24–25 every year. It is a simple mathematical exercise to work out that, even if he spent only one second in each home, he would need several years to complete this trip. Without knowing anything about reindeer propulsion systems, the origin of elves or other such details, we are able to refute the statement on purely logical grounds. As Carl Sagan [7] put it, one second per house would at least explain why no one ever sees him much, but even with relativistic reindeer (travelling at the speed of light) the job is impossible. There is another, similar hypothesis that is held to be true by some people. It states that spacecraft from alien civilizations regularly come from outer space to visit the earth. Can we say anything about this claim? After all, we know very little about alien technology. Sagan [7] applies the Santa Claus analysis to refute this statement.

Example 10.1: Has the earth already been visited?   During some periods there have been many reports of UFOs, so let us assume that we have at least one alien visit on earth every year. Though it is difficult, we could at least try to estimate the current number of advanced technical civilizations in our galaxy. It is usually done using a relationship called the Drake equation. This is a product of a series of numbers and probabilities; for example, the rate at which stars are formed in our galaxy, which is known quite well, the fraction of those stars that have planets, the fraction of planets that are suitable for life, and so on. Needless to say, the estimates become less certain the farther we go into the equation but, without entering into details, a very optimistic estimate of the number of civilizations is 10^6. Now, let us try to calculate how many interstellar spacecraft these civilizations must launch, on average, for one to end up on earth every year.

Say that each civilization sets out on Q interstellar expeditions per year and that each of them reaches one destination. This amounts to $Q \times 10^6$ visits per year, somewhere in the galaxy. Since there are several times 10^{11} stars in the galaxy, there should be at least 10^{10} interesting places to visit. So, in order for one visit a year at any interesting place, such as the earth, we need $Q = 10\,000$ launches per civilization every year. This seems excessive, especially considering the resources needed for such expeditions. Sagan mentions an estimate of the material needed for the spacecraft – they have to be larger than the US Apollo space capsule, let us say – which requires 1% of the stars in the galaxy to be mined.

A possible counter-argument is that we are the object of special attention. Maybe alien anthropologists think that we have just reached a very interesting stage in our technical development? But to imagine that our development is so fascinating is precisely contrary to the idea that there are so many civilizations out there. If there are, the development of technical civilizations must be quite common. If we are looking for an explanation for UFOs, alien visits is probably not the most promising hypothesis.   ∎

Apart from the order-of-magnitude analysis, this thought experiment builds on a useful analogy (comparing the problem to visits by Santa Claus). It also explores the problem from an alternative perspective. The "alien visits" hypothesis uses our perspective, assuming that we are unique enough to be worth visiting. The thought experiment takes the perspective of an alien civilization. They would have to know that we are here in order to visit us. How is that possible without regularly visiting promising worlds?

It should be noted that thought experiments are always based on current knowledge. If we overlook important possibilities, the validity of our conclusions may be affected. Interestingly, Sagan later wrote a novel where an alien civilization contacts us. To make it possible for them to find us, he envisaged a galactic radio surveillance system that could detect our civilization as soon as we started broadcasting television. It was connected throughout the galaxy in a sort of subway system of relativistic wormholes. Assuming the possibility of such a system, the conclusion of his thought experiment may be a different one.

The examples of thought experiments given here are perhaps more elegant than average and we ordinary researchers may think that it is futile to even try to apply the technique. It really is not. The key is to take a playful attitude to the research problems that we encounter. Conceptualize and visualize the claims that you analyze, explore them using different perspectives and scenarios, and ask critical questions. Does the claim lead to unreasonable results under some conditions? What would be required for the claimed effect to occur? As with any other technique, proficiency requires application, so here are a couple of exercises for practice.

Exercise 10.6: In 1633, Galileo was forced by the inquisition to recant his theory that the earth moves around the sun. When doing so, according to legend, he muttered the words "Eppur si muove" (and yet it moves). Is it reasonable to assume that you now hold in your lungs at least one molecule of the air that he breathed out when uttering this rebellious phrase?

Exercise 10.7: In Chapter 7, it was stated that the incidence of caries was greater than average in the town of Caramel. The local newspaper claimed that this was due to the local candy factory expelling byproducts from their production into the ground water. Investigate this claim using a back-of-an-envelope calculation. Assume that the waste is expelled into a river and that a certain fraction of the river's water becomes ground water. Is the newspaper's claim plausible?

Exercise 10.8: "Kaffekask" is a traditional Swedish drink, consisting of vodka and hot coffee. According to an old saying, the perfect proportions are obtained by the following technique. Put a coin at the bottom of a cup. Add strong coffee until you can no longer see the coin and then add vodka until it just becomes visible again. Do you think this method could work? (If you decide to make the actual experiment during office hours, the vodka may be replaced with water.)

10.8 Planning Checklist

Now that you have identified promising hypotheses by discarding the least promising ones using cause-and-effect tables and thought experiments, it is time to devise the actual experiment. In this final section of the chapter we will go through a rough checklist for this part of your investigation.

Firstly, all research begins with a *question*. You should be able to state clearly what you want to learn from the experiment. If you do not have a clear question, you will not obtain a clear answer. If the aim of the experiment is to test an hypothesis you must find a good method for testing it. Which variables are connected with the hypothesis? Which observations could support or disprove it? The answers to these questions will provide clues for a fruitful approach. You may have to build a special apparatus or a dedicated setup of equipment to create the conditions needed for the experiment, or to be able to observe the effect that you are interested in. Each experiment is unique but all involve a question and a method for answering it.

You must choose *response variables* that are relevant for your hypothesis. Sometimes, the effect of interest can be coupled to several different response variables. Be sure to collect *all* the information needed to analyze the experiment and understand the root causes behind the effect. It would be unfortunate to find that relevant information is missing during the analysis.

The response variables are coupled to input variables, or *factors*. Engineers and scientists often take different attitudes to the choice of variables. It is often sufficient for engineers to optimize a system without understanding its fundamentals, so the input variables may be any "control knobs" that are practically useful. Scientists aim to understand and explain. Their input variables must often be more carefully chosen because they should be coupled to *root causes*. This is discussed further in the next chapter.

The next step is to choose an *experimental design*. It could be one of the standard designs described in the last chapter but, as established in Chapter 6, you may need to devise a unique measurement strategy for your specific problem. Approaches generally vary according to the nature of the factors. If they are categorical, the experiment tends to focus on the comparison of two or more conditions. Randomized controlled experiments, crossover and factorial experiments are common examples. With continuous, numerical variables, experimenters tend to sweep variables or use response surface designs. The design must be aligned both with the objective of the experiment and the resources at hand, because experiments can be expensive. The number of experimental treatments or measurements may be limited by the supply of raw materials, time constraints or other resources. If so, you should strive to obtain the data that is most pertinent to the research question. When designing your experiment it is also important to consider the need for a control case to provide a reference level for the effects of your treatments. Finally, you need to consider noise and background variables. Noise can be due to variations in experimental settings, environmental factors, the detection system and so forth. They are attended to by replication and randomization. Background variables are less random in nature and can cause drift or other false patterns in the response. Such effects are often avoided by replication and blocking. As previously mentioned, there may be other methods for handling noise and background factors that are more suited to your needs.

Once you have devised an experiment with responses and factors, it is time to decide on the levels that the factors should be set to in the experiment. These define the *ranges* of the input variables. It is often useful to vary the factors in wide ranges, as this makes their effects easier to detect. Be careful, however, not to extend the range to the point that the system under study fails to function normally. If the range is too small, on the other hand, you may not be able to detect an effect, even if one exists. Appropriate factor levels can usually be identified in an initial screening phase through variation around a known baseline setting. It is always dangerous to give general guidelines but, as a rough rule of thumb, if your response varies by a factor of two compared to the baseline condition, the effect is readily detectable.

Before collecting data you also need to plan *how to collect the data*. Do you, for example, know the capability of your measurement system? If not, you should make a measurement system analysis, which will be described further in the next chapter. We also learned in this chapter that it is important to determine the scope of your study. It may be difficult to answer every aspect of your research question in a single experiment. If so, you should break the problem down into several parts and attend to them individually. If you are using design of experiments, it is useful to make sequential experiments: starting with a screening design and then adding treatments as you proceed, maybe to obtain a response surface design. Never use the majority of your resources on your initial tests; save them to the later stages when you have found and solved the bugs in your setup.

The above activities can be summarized in the following checklist:

- Clearly state the question
- Select appropriate responses
- Select useful factors
- Design the experiment
- Determine factor ranges
- Plan the data collection.

Experiments often fail because items on this checklist have been overlooked. A factor may be varied in too narrow or too wide a range, the measurement system may be too poor, or the experiment may not have been randomized. Furthermore, as a general rule, things can and will go wrong. It is, therefore, useful to formulate "Plan B" at an early stage. As Dwight D. Eisenhower put it, "I have always found that plans are useless, but planning is indispensable."

Last, but not least, when planning and conducting an experiment it is vitally important to continuously document your thoughts and observations. It may be weeks or months before you analyze the data and your notebook will become a valuable asset when your memory fails to recall details of discussions and events in the laboratory.

10.9 Summary

- The novelty requirement can be fulfilled by showing originality with respect to: data; methods and techniques; research questions; research areas; or analysis and synthesis.
- Why-analysis can be used to explore the boundaries of knowledge and identify promising research problems. It helps us to penetrate layers of descriptive knowledge to a level where our understanding of a problem is finally challenged.
- Since research problems often are complex it is important to break them down into subproblems that can be handled, as well as setting a realistic scope for the research.
- Ishikawa diagrams allow us to summarize important relationships between an effect and its potential causes in a process or phenomenon.
- Cause-and-effect tables can be used to explore how we expect a potential cause to be coupled to its effect. They are thereby useful for developing ideas for experiments.
- Process diagrams are tools for visualizing steps in a process. They are useful in the formulation of research questions and the development of ideas for experiments.

- Thought experiments are used for evaluating hypotheses by exploring their logical implications. If an idea leads to paradoxes it is not generally applicable and may be discarded.
- The tools above are useful for generating, evaluating and ranking research hypotheses. When a clear research question has been stated the planning phase continues by identifying appropriate responses and useful factors, designing an experiment, determining factor ranges, and planning the data collection.

Answers for Exercises

10.1 Experiment 1 is based on a new research question (Can insects use the Milky Way for orientation?), provides new data (on this particular beetle under specific conditions) and uses a new method (placing the beetles in a planetarium). Experiment 2 is based on a new research question (Do hot combustion products affect the lift-off length?), provides new data (lift-off lengths in a particular optical engine under specific conditions) and uses a new technique (comparing a symmetric with an asymmetric nozzle).

10.2–10.5 Left as exercises for the reader.

10.6 Here is one possible approach. Assume that the molecules in this breath are now evenly mixed in the atmosphere. First estimate the fraction of the atmosphere that these molecules make up. There is 10^5 kg air per square meter earth (the air pressure is 10^5 N/m^2). The earth's diameter is 10^7 m (the circumference is, by definition, 40 000 km) and the surface area is roughly the diameter squared, or 10^{14} m^2, giving us 10^{19} kg of air on earth. A breath is 10^{-3} m^3 and the air density is 1 kg/m^3, so Galileo's famous breath makes up the fraction $10^{-3}/10^{19} = 10^{-22}$ of the earth's atmosphere. Hence, you must breath in 10^{22} molecules for one of these particular molecules to end up in your lungs. You currently hold 10^{-3} kg of air in your lungs with a molecular weight of 10^1 kg/kmol, amounting to 10^{-1} mol or roughly 10^{23} molecules. Amazingly, it is fair to assume that your lungs now contain some of the molecules he breathed out when (if) he uttered the famous words.

10.7 Here is one possible approach. A conservative guesstimate is that the water can cause caries if it has 1% of the sugar concentration of a commercial non-diet carbonated drink (about 10%). A quick internet search gives that the water flow of a moderately sized river is about 50 000 kg/s. To produce the required sugar concentration of 0.001 the factory must expel 50 kg/s of sugar. This amounts to 10^9 kg/year, which is at least one order of magnitude greater than the annual sugar production of a country like Sweden. The newspaper's hypothesis is not plausible.

10.8 Here is one approach. The reason that coffee appears dark is that it absorbs white light coming into the cup from the top. (The effect could potentially have been due to scattering instead of absorption, but then the liquid would appear white as milk.) When most photons are absorbed on their way from the surface to the coin and back, the coin is no longer visible. Dilution by a clear liquid will not change the total number of absorbing "coffee molecules" in the light path and thereby cannot render the coin visible again. The recipe is probably a joke made by someone who likes a strong cup of kaffekask.

References

1. Dacke, M., Nilsson, D.-E., Scholtz, C.H., *et al.* (2003) Animal Behaviour: Insect Orientation to Polarized Moonlight. *Nature*, **424**, 33.
2. Dacke, M., Baird, E., Byrne, M., and Warrant, E.J. (2010) First Evidence of Star Orientation in Insects. Paper presented at: International Conference of Neuroethology, Salamanca, Spain.
3. Chartier, C., Aronsson, U., Andersson, Ö., *et al.* (2011) Influence of Jet-Jet Interactions on the Lift-Off Length in an Optical Heavy-Duty Diesel Engine.Submitted to *Int J Engine Research.*
4. Chartier, C. (2012) *Spray Processes in Optical Diesel Engines: Air Entrainment and Emissions* [Ph.D. thesis], Lund: Lund University.
5. Gamow, G. (1966) *Thirty Years That Shook Physics: the Story of Quantum Theory*, Doubleday & Co., New York.
6. Ishikawa, K. (1990) *Introduction to Quality Control*, Chapman & Hall, London.
7. Sagan, C. (1973) *The Cosmic Connection: an Extraterrestrial Perspective*, Dell Publishing, New York.

11 Phase II: Data Collection

Thinking is more interesting than knowing, but not as interesting as looking.

—J.W. von Goethe

In this chapter we will discuss two aspects of the data collection process. The first has to do with *what* we measure and this discussion builds on the contents of the last chapter. We must choose variables that are connected with our research question in a meaningful way. If they are not, it is difficult to draw meaningful conclusions from the data. The other aspect has to do with *how* we measure. It is important to make sure that the data are of sufficient quality to support the conclusions that we wish to draw and we ensure this by using proper measurement systems and procedures. We are going to discuss how to develop a measurement system, how to identify sources contributing to measurement uncertainty and how to quantify the uncertainty. Finally, we will go into the benefits of making a data collection plan.

11.1 Generating Understanding from Data

Every day we hear claims that are based on observations. Some of these observations provide hard data that can be expressed exactly in numbers, while others provide soft data that are not easily quantified. For example, saying that the weather is cold could be based on hard data from an outside thermometer, but saying that the neighbor looks happy today is certainly based on soft data.

The famous physicist Lord Kelvin said that it is important for scientists to be able to measure what they are speaking of because "when you cannot express it in numbers, your knowledge is of a meagre kind". This statement is often criticized because it plays down the value of soft data. It must be acknowledged that soft data are very useful in many situations. It is also true that quantitative data are not the same thing as meaningful knowledge. To realize this we only have to recall the methodical gardener who counted his apples in Chapter 2. However unique the information, knowing that there are 1493 apples in his garden is not terribly interesting, unless this is stated in the light of a meaningful problem. Data must answer a well-posed question in an unambiguous way to be useful; formulating that question was the central topic of the last chapter.

The point of Kelvin's statement is that, to provide an unambiguous answer to a research question, observations must be quantified at some level. Research requires a high degree of inter-observer correlation, meaning that other researchers must be able to repeat your experiment and come to the same result. Scientific conclusions cannot be based on subjective opinions. If you are unable to quantify responses in an experiment your data cannot be used

---

Experiment!: Planning, Implementing and Interpreting, First Edition. Öivind Andersson.
© 2012 John Wiley & Sons, Ltd. Published 2012 by John Wiley & Sons, Ltd.

to support any statement about effects. This, indeed, is a meagre kind of knowledge. When quantitative data are not meaningful, the problem always lies with the question and not with the data.

Experimental equipment often has a number of "control knobs". This is especially common in applied research. Machines, instruments and industrial processes have actuators and other components that are controlled using these knobs. As they are intentionally easy to manipulate, it is often tempting for an experimenter to use these control knobs as factors in an experiment. On the other hand, it is important to think carefully about how the experimental factors are connected to your research problem. To be able to explain your observations as well as possible you should investigate root causes; the "control knob variables" seldom belong to this category. Although a chemical plant, for example, is controlled using valves, valve settings say little about what is actually happening in the chemical process. Such control variables may provide descriptive knowledge about how a process may be tuned but offer little in the way of explanations.

As repeatedly stated in this book, science aims to explain rather than just describe, and scientists therefore need to focus on explanatory variables. In the case of a chemical plant, it is probably more interesting to know something about concentrations of substances or local temperatures than about the settings of valves. An engine developer might be interested in the throttle position of an engine, but an engine researcher is likely to be more interested in the corresponding air mass-flow into the engine, or the level of turbulence that it produces in the cylinder. A developer of drugs may be interested in the concentration of a substance in a pill, whereas a medical scientist is more interested in how the same substance acts in the body.

Understanding how something works requires more direct measures than those needed to control it. To understand and explain you must connect the output from your experiment to cause variables. These are often more difficult to identify and significantly more difficult to manipulate than the control knobs on the apparatus in your laboratory. If you find that your research question involves control variables rather than cause variables, consider revisiting the planning phase of the last chapter before proceeding with data collection.

As described in Figure 11.1, data collection is the research phase that takes place between the planning and the analysis of an experiment. The smaller black arrow indicates that it is sometimes necessary to return to the planning stage before finishing the data collection. This could be due to unexpected results that change the focus of the study, or to practical problems. In the beginning many Ph.D. students consider data collection to be the most important activity in research. Data are necessary to draw conclusions but the actual collection of data often makes up a relatively minor part of a research project. In many respects, it is the most technical part of the scientific process. In the following sections we will discuss the central steps in the data collection phase. They treat the development of measurement systems, how to assess the accuracy and precision of such systems, and how improve them. Firstly, however, we will discuss an inherent property of all measurements – uncertainty.

11.2 Measurement Uncertainty

When you step onto the bathroom scales in the morning, it translates the gravitational force, F, that you exert on it into a readout, which is your weight, W. The relation between them can be expressed as:

$$W = a + bF + \varepsilon \tag{11.1}$$

Figure 11.1
The three phases of research. During the data collection phase it may be necessary to revisit the planning phase, due to practical problems or unexpected results.

where a represents an offset, b is a proportionality constant (or slope), and ε is the random measurement error. If the scales are well calibrated the readout should be very close to zero when there is no weight on the scales, allowing for some random error. This means that a should be zero. The slope b should also be tuned so that correct readouts are produced when different weights are put on the scales. If either of these two parameters is off we say that there is *bias* in the readout. If ε is large we say that the readout suffers from *noise*.

This equation is quite a simplified description of the measurement chain, as the force is not directly translated into readout. In mechanical scales, the force of your body acts on levers, which extend a spring that turns a dial through a rack and pinion arrangement. Each of these components contributes both to the bias and the noise in the readout, and adds to the overall measurement uncertainty. If you want to improve the quality of your data, you must identify the components that contribute the most to the uncertainty and focus your efforts there.

Measurement uncertainty generally comes from a range of sources. Bias is not only due to poor calibration. It can result from drift; for example, if parts in an instrument become worn over time. Environmental factors may also affect the measurements, if humidity, ambient temperature or something similar affects the properties of the components in an instrument or of the measurement object. The measurement procedure is another source. For example, the readout from the scales will depend on if you weigh yourself before or after breakfast, your amount of clothing and so forth. If you do not settle on a specified procedure for weighing, such factors may add both bias and noise to the data. To understand the sources of uncertainty you must consider measuring instruments as well as measurement objects and measurement procedures, and investigate how these are affected by operator skill, calibration procedures, environmental factors and so on. In other words, measurement uncertainty does not only come from the measuring instrument. It comes from the whole measurement system, including the operator, the test object and the instrument.

When discussing measurement data it is common to divide the error into a random and a systematic part. Random error produces randomly different results from repeated measurements and averaging over several measurements tends to reduce its effect. The random error is coupled to the measurement precision, or the noise as it was called above. Systematic error means that each repeated measurement at a certain condition is affected

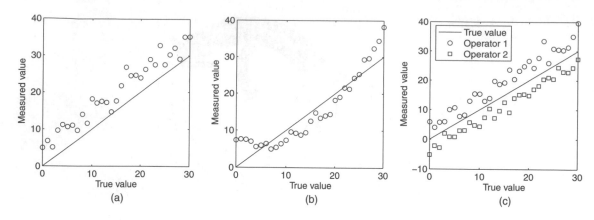

Figure 11.2

(a) Equal bias over the entire range of values; (b) unequal bias, where the offset is a function of the measured value; and (c) operator bias, where the offset is a function of the operator.

by the same influence. This situation is not improved by repeating measurements. It results in bias and is coupled to the measurement accuracy. A systematic error can simply be a constant offset over the entire range of values but, as illustrated in Figure 11.2, bias can also be more complex.

Systematic errors must be handled differently than random errors. Calibration of instruments is one important method. We may also know about some background effects and make corrections for them. For example, if we measure the length of objects that are known to expand with temperature, we could employ a temperature correction in the measurement procedure. To avoid uncertainty depending on operator skill or personal judgment it is also useful to formulate a standardized procedure for measurement and make sure that the data are collected according to it. A useful way to monitor the measurement quality is to run regular stability checks, as will be described at the end of this chapter. Repeating a certain measurement regularly, for example once a day, makes it possible to detect drift or sudden changes in the performance of the measurement system.

It is important to note that, although systematic and random errors are defined differently, it can sometimes be difficult to distinguish between them. The temperature expansion of the objects above, for example, is a systematic effect. On the other hand, if this effect is not known, repeated measurements of the same object would yield unexplained differences in the results if the temperature were not constant. This would be interpreted as random noise in the measurements.

11.3 Developing a Measurement System

Experiments are devised to answer unique questions. For this reason, experimenters must often develop unique setups and unique measurement systems to obtain relevant data. It is of course difficult to give general advice that is useful in all possible situations across several disciplines, but there are a number of general problems that everyone setting up a measurement system has to consider.

Firstly, it is important to note that the measurement system includes the measuring instrument, the measurement object and the operator, and so is very much part of the experiment itself. In many of the previous examples in this book we have seen that the measurement system is co-developed with the experiment. In Chapter 6 we saw how Millikan developed the balanced drop method by trying to avoid the uncertainties inherent in the charged cloud method. In Experiment 1 in the last chapter the wooden arena for the dung beetles was developed to avoid the beetles using the camera stand for orientation. In Chapter 4 we saw how central the details of the measurements were to the successful implementation of Galileo's experiment on the inclined plane. The method for measuring time did not only have to be accurate in itself but also practically suited to the experiment. Keeping an internal rhythm by singing a song actually provided better time measurements than using the swings of a "sophisticated" pendulum, which would require the coordination of two external cues. The channel in the inclined plane was "straight, smooth and polished" to improve the data quality. Settle's reproduction of the experiment showed the importance of allowing the operator a few practice runs before measurement. Clearly, the development of a successful experiment, in many respects, is a process of finding a measurement system that is suited for the problem of interest.

Sometimes, research students are reluctant to engage in practical aspects such as developing or building experimental setups, although this work is often inseparable from the research task itself. It is just as important to research as making a literature study or analyzing data. The sooner we engage in the practicalities of experimentation the sooner we will become productive researchers. Taking an experiment from start to finish often requires that we develop new competencies. Perhaps we will need to learn a bit of electronics, a new programming language, something about materials, or a new statistical technique. Similarly, since each experiment is based on a unique research question, we cannot expect our laboratory to be furnished with equipment that is ready to produce results that are relevant to our problem. Pottering about with equipment and solving practical problems is an important part of what researchers do. How successful would Millikan have been if he had decided that it was someone else's job to develop the balanced drop method? How far would Mendel have come if he had not engaged in collecting and drying his peas or weeding the flowerbeds? Even if it is wise to leave some activities to others, such as manufacturing equipment, it is easier to obtain useful results from an experiment if you are engaged at some level during all steps from idea to conclusion.

When developing a measurement system, the first step is to determine what to measure. This is often straightforward, as the choice of variables is likely to be a natural consequence of the experimental idea. As previously mentioned, we should strive to use cause variables rather than control variables, as they make it easier to explain our results. We also need to form an idea of how to steer clear of background variables. In the last chapter we used the Ishikawa diagram to identify cause variables. As this diagram points to several variables that may influence the result it is useful also for identifying potential background variables in an experiment. The final purpose of identifying them is to develop strategies for avoiding their influence.

Exercise 11.1: Imagine that you are developing a measurement system to investigate if the dung beetles in Experiment 1 (Chapter 10) can orient to the Milky Way. Look at the Ishikawa diagram in Figure 10.6. Does it point to potential background variables that could spoil your analysis?

Example 11.1: During the early dung beetle experiments described in Chapter 10, the arena was a three-meter wide, circular area of flattened and leveled sand, enclosed within a circular cloth wall. The beetles were filmed using a video camera suspended from a wooden gantry above the center of the arena. During nighttime experiments the camera used a "Night Shot" facility, filming the arena in infrared light. A new setup had to be developed for the planetarium experiment. This was partly because the space available under the planetarium dome was limited, but also because the beetles could potentially use the camera stand as a landmark for orientation. The team had to develop a new method that did not involve a camera.

They constructed a two-meter wide, circular wooden arena that would completely eliminate the potential influence of cameras or other objects that could serve as landmarks. The beetles were to see nothing but the planetarium dome during the experiments. To achieve this, the arena was elevated one meter to allow the experimenter to sit beneath it. A small, circular trap door at the center of the arena could be removed from below to admit the beetle and its ball to the surface. The door was then inserted back into its original position in the floor of the arena, from where the beetle could begin to roll.

Instead of filming, the team assessed the straightness of the beetle's rolling path by measuring the rolling time. A timer was started when the beetle left the center of the arena. It was stopped at the sound produced when the ball fell over the edge into a collection trough located around the perimeter of the arena. With this measurement system the team could distinguish between situations where the beetle could orient itself and where it could not. A beetle would typically complete a perfectly straight rolling path in 20 seconds, while a convoluted path could exceed five minutes in duration. ■

Another important property of a measurement system is that it must be practically useful. This may lead to trade-offs, as resources are always limited. If the measurements become too complicated, this may limit the amount of data that you can collect. On the other hand, the research task dictates the amount of data that you must collect to draw meaningful conclusions. As a result you may have to choose a simplified measurement strategy to achieve your goal. Let us use a simple example to discuss why simplifications do not necessarily decrease the quality of our conclusions.

As discussed in Chapter 4, researchers are frequently more interested in patterns and regularities than absolute units. An hypothesis could, for instance, suggest that lack of a certain enzyme causes a certain disease. This hypothesis would be discredited if a relevant set of data did not show any correlation between the level of the enzyme and the symptoms of the disease. Neither of these two variables would have to be measured in absolute units to draw the conclusion so long as the *trends* in the data were reliable enough to calculate the correlation (see the section on correlation in Chapter 8). You could have access to a measurement technique that yields the enzyme concentration in relative units. Turning them into absolute concentrations may require a substantial effort and, since absolute units are not needed to test the hypothesis, this effort would provide no additional value to your experiment.

Another practical aspect of measurements is that they often disturb the state of the measurement object to some degree. When measuring a temperature, for example, the thermometer must be in thermal equilibrium with the measurement object to provide a correct reading. This means that heat is transferred between the two and that the temperature is affected. The effect will be small if we measure the water temperature in the ocean, but substantial if we measure in a test tube. If we wish to measure the temperature in a flame by putting a thermometer into it we will disturb the flow in addition to cooling it. It is simply not the same flame with a thermometer in it. There are parallels to these scenarios in most measurement situations. To measure is to affect. In order to obtain representative data it is important to find ways to minimize this disturbance.

To develop a reliable measurement system we need to attend to the whole measurement chain. Information is fed into the measurement system and is transformed in various ways before it comes out as data. This information flow was exemplified above with the levers and springs of a bathroom scale, but a measurement chain – short or long – exists in every measurement situation. Even when measuring something as simple as the length of a rope a number of influences affect the result. The method for aligning the rope with the measuring tape is one, the method for reading the tape is another, and the expansion of the tape with temperature is yet another. A measurement can be seen as a process where each step contributes to the error. One way to identify the parts that are most critical to the reliability is to make an Ishikawa diagram. This diagram will be different from the one drawn in the last chapter because it focuses only on the measurement process. As the measurement represents a flow of information from input to output, the arrows of the diagram can be added from left to right in the order that the influences appear in the process. It helps to think in terms of the "5 Ms" when searching for all the relevant influences (man, machine, method, material, and measurement). Do not forget to consider environmental factors too.

Example 11.2: A measurement on the sand arena consisted in releasing a dung beetle and capturing how it rolled its ball on film. Some influences on the outcome of this measurement chain are summarized in the Ishikawa diagram in Figure 11.3. We see that a number of practical considerations are necessary to avoid systematic and random errors. It is, for instance, important to handle the beetles in the same way during each measurement. Their

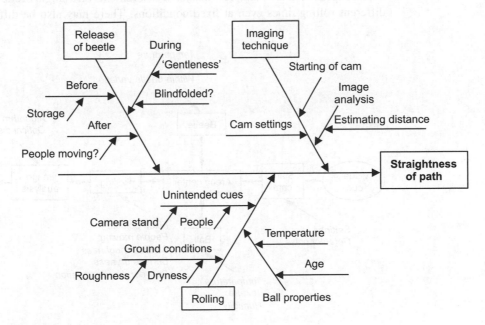

Figure 11.3
Ishikawa diagram of the measurement system for the beetle experiment.

performance could be affected if they are allowed visual cues before release, compared to if they are kept in a dark place ("blindfolded") or if the gentleness of the release varies. ∎

When the potential influences have been identified it is time to consider each influence in closer detail. It is convenient to document them in a process diagram. Close your eyes and visualize the steps through which the input is turned into an output in the measurement chain. List influences from all parts of the measurement system that add to the uncertainty in each step, including the operator, the measurement object and the environment. They may result in both systematic and random error. After taking this mental walk through the measurement process, take a sheet of paper and draw the process diagram. A process diagram based on the Ishikawa diagram above is shown in Figure 11.4, where oblique arrows indicate sources of uncertainty.

The purpose of making these diagrams is to identify as many potential sources of uncertainty as possible. As a final step it is useful to summarize and categorize these sources according to how they can be handled. This will be done in what we will call an *input–output diagram*. At first sight we may think that the measurement process only has one input and one output, but looking at the process diagram we realize that this is not quite true. The output is the information that you obtain from your measurements. In the beetle experiment this is the straightness of the path and in the diesel engine experiment (Experiment 2 in the last chapter) it is the lift-off length. The inputs, on the other hand, can be divided into three different categories. Hopefully, the treatment that you as experimenter exert on the system under study is the most important input, but background factors and noises also affect the output. Background factors lead to systematic errors and noises decrease the measurement precision. The noises can, in turn, be divided into internal and external noises. The internal noises arise from the measurement object itself. In the beetle experiment, for example, repeated measurements on a single beetle will give slightly different rolling times even at fixed conditions. There may also be differences between

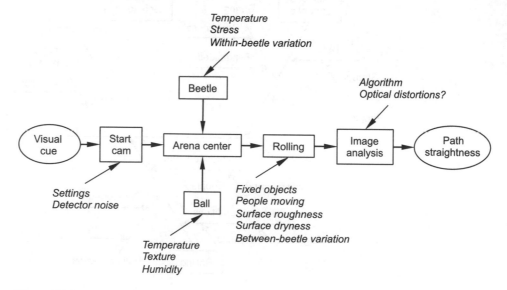

Figure 11.4
Process diagram based on the Ishikawa diagram in Figure 11.3.

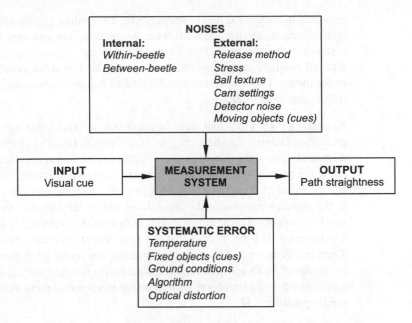

Figure 11.5
Input–output diagram for the beetle experiment, listing the error sources by category.

the beetles. This could seem like a systematic error but it will appear as noise in the measurements, at least if differences between the individuals are not assessed before the experiment. External noises arise outside the measurement object. They could be due to varying environmental conditions or to the operator failing to repeat settings exactly when a measurement is replicated. Figure 11.5 shows an input–output diagram based on the process diagram in Figure 11.4. The measurement system is represented by a box at the center; the inputs and output are connected to it by arrows.

Using this diagram we can now formulate countermeasures to avoid or at least decrease the effects of the undesired influences. Systematic errors are often handled by calibration or other types of corrections. Randomization is another common method to mitigate their influence, especially for influences that cannot be corrected for. Since noises are random influences it is not possible to correct for them but they can often be decreased. External noises can be decreased for example by isolating the experimental setup from environmental influences, such as vibrations and temperature fluctuations. If the measurement procedure is found to be a source of noise it can be amended by defining the measurement procedure more strictly and making sure that the operator adheres to it. Automation is another common method, since computers or machines tend to perform repetitive operations more consistently than humans. It is more difficult to decrease the influence of internal noises. Often, the only thing to do is to quantify them.

In some research fields, noisy signals are filtered. The interesting variations in a signal may occur over a longer timescale than the noise. Removing higher frequencies may then decrease the noise and make further analysis possible. In such cases it is important to filter with care in order not to lose useful information.

Example 11.3: In the beetle experiment, the research team suspected that the rolling speed depended on the temperature of both the beetle and the dung ball. The velocity was expected to increase at higher temperature. To obtain a quantitative relationship describing this influence, beetles and balls were warmed to temperatures in the range 17–25°C in a steam-bath chamber before timing the straight rolling trajectories in open conditions with all available cues. This relationship was used to avoid systematic errors in the actual experiments. The temperatures of the air and ball were measured in the field to correct the rolling durations. ■

Example 11.4: Vibrations were an important external noise factor in Experiment 2 in the preceding chapter (the diesel engine experiment). The lift-off length was measured as the distance between the injector and the flames on the fuel sprays. It was therefore essential to know the precise location of the injector in each movie frame. Two problems needed to be solved. Firstly, the engine was moving during a combustion cycle, leading to slight changes in the injector position in the images. Secondly, the injector was not directly visible. A specific tracking algorithm was therefore developed to indirectly locate the injector in each movie frame. It used the fact that the four fuel sprays were visible close to the injector. Their positions were determined by tracking the signal level along a circle drawn around the center of the image. The brightness peaked at four angular locations on the circle. These locations provided four coordinates, and the intersecting point between them was the actual nozzle position. ■

> **Exercise 11.2:** Identify sources of uncertainty in an experiment that is relevant to your own research. Make an Ishikawa diagram of the measurement process and transfer the influences to a process diagram that indicates the types of uncertainties connected with each source. Summarize the uncertainties in an input–output diagram, with internal noises, external noises, and systematic errors listed separately. Finally, think about how you could improve your measurement system.

11.4 Measurement System Analysis

In the preceding section we focused on identifying the sources of uncertainty. We also discussed the ways in which different types of uncertainties can be handled. The time has now come to quantify how great the uncertainty is. Before we have determined the accuracy and precision of our measurement system we do not know if it is sufficient for our purpose.

As the *precision* reflects a random variation between measurements, it is obtained by replicating the same measurement many times. These replicates should be taken at different times to include as many sources of noise as possible. External noises, especially, tend to vary over time. They arise through small variations in the environment, difficulties in repeating experimental settings and so on. This means that consecutive measurements tend to be more similar to each other than measurements made at different occasions, so they underestimate the effects of external noises. One of the better ways to assess the precision is to run regular stability checks during the measurements – a procedure that will be described further at the end of the next section.

Let us say that we are interested in the precision of the measurements with the bathroom scales above. This is obtained by weighing the same person at a number of different occasions. The person's weight will of course fluctuate from day to day depending on food intake and so on, so it may be tempting to use a constant weight instead of a person.

By doing that we would exclude the natural variation in the person's weight from the measurement system. This may or may not be appropriate, depending on which part of the uncertainty we are interested in. We should be aware that excluding this variation leads to an underestimate of the total variation in the measurements.

Imagine that the purpose of our measurements is to test the weight-reducing effect of a diet. This effect is likely to manifest itself as a trend over several weeks. In that case the day-to-day fluctuations must be included when assessing the precision, because it tells us how large the effect must be to stand out from the noise. If the purpose is to measure the day-to-day fluctuations, on the other hand, this variation should of course not be included in the assessment. In the end, it is a matter of judgment which sources of variation to include.

One of the most common methods for improving the precision in a measured value is to average over several measurements. Although this does not decrease the variation in the data, it provides a more reliable estimate of the true value. The underlying reason is the central limit effect (Chapter 7). It is important to distinguish between the precision in the raw data and in the estimated value.

Determining the *accuracy* involves comparisons. For example, if you want to measure the concentration of a certain substance, the accuracy is obtained by measuring a sample of known concentration. The offset between the measured value and the true one is a measure of the accuracy. This procedure is used when calibrating a measurement system, where the accuracy is improved by adding a calibration constant to the measured value to eliminate the offset. Note that, whereas increasing the sample size may increase the precision in the measured value, it has no effect on the accuracy. The difference between precision and accuracy is illustrated in Figure 11.6.

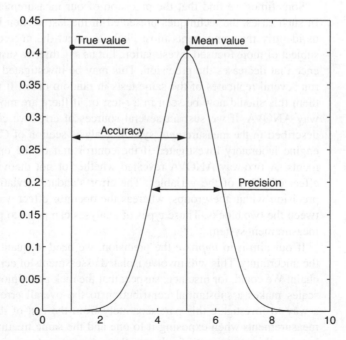

Figure 11.6
The normal distribution represents the measurements. The accuracy is given by the distance between the distribution mean and the true value. The precision is given by the variation in the measurements.

It is not always possible to produce a well-defined standard for calibration in the laboratory. If you are interested in measuring the concentration of a short-lived radical in a flame, for example, you are facing the practical problem that flame radicals may cease to exist only nanoseconds after being formed. This makes it difficult to produce a sample of known concentration. In such cases it is strictly impossible to establish the accuracy in absolute units, but corrections may still be necessary to remove bias and to improve the *relative* accuracy. This can be illustrated by the beetle experiment, which is an example of a study based on relative comparisons. Only relative numbers are needed to differ between situations where the beetle can and cannot orient itself. The rolling time is merely an indirect measure of how straight the rolling path is. As long as the rolling times for straight and convoluted paths are substantially different, this indirect measure is fully sufficient to make an unambiguous distinction between the two situations. As a matter of fact, absolute units are not at all interesting in this context. We could make the distinction equally well using a clock going at the wrong rate. But even though the experiment was based on relative data, Example 11.3 showed that the rolling times needed a temperature correction to remove a bias that could otherwise have made the relative comparison less accurate. Even in cases where measurements cannot be calibrated in absolute units it is important to consider the relative accuracy.

When we have established the precision, our next question is if it is sufficient for the purpose of the experiment. The precision determines how small a difference you can detect between two sets of data. As previously stated, the precision in an averaged value is improved by increasing the sample size. We can determine the sample size required for detecting a certain difference using a technique called *power analysis*. It was discussed in a separate section of Chapter 8.

Sometimes we find that the precision of our measurement system must be improved. In such cases, the techniques presented in the last section become helpful. They helped us identify the components most likely to affect the uncertainty. These will now be the subject of more focused investigation. Let us say that we suspect the operator has an influence that decreases the precision. This may be investigated by letting different operators run several replicates of the same tests in random order. If there are differences between them this should now be seen in a *t*-test or, if there are more than two operators, a one-way ANOVA. If we suspect several sources of error we could use the type of analysis described in the measurement system analysis section of Chapter 8. In that example, an engine laboratory investigated if the control unit or the operator affected the measurements. A two-way ANOVA revealed whether or not there was a statistically significant effect of either of the variables. The error standard deviation gave us the measurement precision within the groups, whereas the operator effect was given by the difference between the two blocks. These types of analyses can point to potential improvements of the measurement system.

If our aim is to *improve* the precision, we need to quantify individual components of the uncertainty. This will involve isolated assessments of certain parts in the measurement chain. We could, for instance, suspect that the rack and pinion mechanism of the bathroom scales makes a substantial contribution to the overall error. To find out, we must find a way to investigate this part separately from the rest of the scales. By taking repeated measurements when exposing it to one and the same treatment we will get a measure of how much uncertainty it feeds into the measurement process. When we find components that are responsible for a large portion of the overall variation we know where to focus our improvement efforts.

There are different ways to *express* measurement precision. As explained in Chapter 7, the standard deviation is one of the more convenient measures of the amount of variability in a data set. Firstly, if the error is normally distributed the standard deviation can be used to identify outliers. This is because it is very unlikely to find values more than three standard deviations from the mean of a normally distributed sample. It is also convenient that the standard deviation has the same unit as the measured value. If we instead want to express the uncertainty of values that are averaged over a sample, the *standard error of the mean* is a better measure than the standard deviation. You may, for example, average your weight readings over a whole week to obtain a more reliable value. It is not the variation in the data set that is important in this case but how well the population mean is estimated. We obtain the standard error of the sample mean, $\bar{x}$, by dividing the sample standard deviation, s, by the square root of the sample size, n:

$$s_{\bar{x}} = \frac{s}{\sqrt{n}} \tag{11.2}$$

Here, s is an estimate of the population standard deviation, σ, and the reason that this formula estimates the error in the sample mean is given by the central limit theorem (Chapter 7). Yet another way to express the uncertainty is to use a confidence interval. Remember that these rules are not strict, so it is advisable to always state which error estimate you are using.

If the output value from an experiment is calculated from several measured values, evaluating the uncertainty can be considerably more complex. The total variation can then be calculated from the measurement variation using what is called *error propagation analysis*. Say that the value of interest, f, is a function of the two independent, measured variables, x_1 and x_2. If the error variances of these variables are σ_1^2 and σ_2^2, respectively, the error variance of $f(x_1, x_2)$ is approximately given by:

$$\sigma_f^2 \approx \left(\frac{\partial f}{\partial x_1}\right)^2 \sigma_1^2 + \left(\frac{\partial f}{\partial x_2}\right)^2 \sigma_2^2 \tag{11.3}$$

The terms in the brackets are partial derivatives, meaning derivatives of f with respect to one of the variables when the other is held constant. This formula can be extended to any number of independent variables. To make the calculation more concrete, consider a situation where we are interested in determining the volume, V, of a cylinder from its measured diameter, d, and height, h. Call the error variances of these measures σ_d^2 and σ_h^2, respectively. The volume is given by:

$$V = \frac{\pi}{4}d^2 h \tag{11.4}$$

The partial derivative of V with respect to d is:

$$\frac{\partial V}{\partial d} = \frac{\pi}{2}dh = \frac{2V}{d} \tag{11.5}$$

Similarly, the partial derivative with respect to h is:

$$\frac{\partial V}{\partial h} = \frac{\pi}{4}d^2 = \frac{V}{h} \tag{11.6}$$

This means that the error variance in V is approximately given by:

$$\sigma_V^2 \approx \left(\frac{2V}{d}\right)^2 \sigma_d^2 + \left(\frac{V}{h}\right)^2 \sigma_h^2 \qquad (11.7)$$

Which of the two error variances on the right contributes most to the overall uncertainty clearly depends on the values of d and h.

Exercise 11.3: Imagine that an experiment requires you to prepare a large number of samples of a certain fixed volume. The pieces of sample are cut from bars with a mean diameter of 1.7 mm. To obtain a volume of 4.3 mm^3, they are cut in lengths of 1.9 mm. Upon closer inspection you find that the standard deviation of the diameters is 0.04 mm and the standard deviation of the lengths is 0.07 mm. Which of these standard deviations should be decreased if we want to decrease the variation in the volume?

11.5   The Data Collection Plan

We have now decided what to measure, developed a measurement system for it, corrected for bias, and identified and quantified important sources of error. Before proceeding with the actual measurements we should now formulate a data collection plan. As this plan should be based on the particulars of your measurement system it will be unique for each experiment. It is simply a written description of how the measurements are to be carried out in order to obtain representative data of good quality. As the input–output diagram lists noises and background factors it is a good starting point for the plan. The purpose is to specify practical procedures for diminishing such influences.

Background factors can sometimes be controlled, for example by isolating the measurement system from them. In other cases we can make corrections, for example by calibration. If they vary predictably we can attend to them by blocking. If they are expected to drift slowly with time we may have to calibrate at regular intervals during the measurements. It is more problematic if they vary quickly in an unpredictable way. In such situations we may measure the background factor and record it as a covariate, which can then be used for corrections of the data during the analysis phase. Remember that some background factors will remain unknown; this is why it is recommended that the run order of the measurements is randomized.

Although external noises are random effects they can often be reduced. Example 11.4 showed one example of how the influence of external noise in the form of vibrations could be decreased. The measurement procedure is often an important source of external noise, since there is always some variation in how the experimenter makes settings or creates the conditions for the measurements. For example, the measurement system analysis section in Chapter 8 showed that the two operators obtained different results from what should have been identical experimental conditions. Such differences are often due to differences in the subjective understanding of how the measurements are to be made. In the example, the difference could be due to one operator running the engine warm for a longer time before taking the measurement. We cannot blame people for making personal interpretations if we did not specify the measurement procedure it in detail. For this reason, a detailed and standardized procedure for measurement is a central part of the data collection plan. Even if only you make the measurements, you must decide on how to proceed. Otherwise, measurements made when you are tired towards the end of the day may be quite different from those made in the morning.

Example 11.5: In the diesel engine experiment, the lift-off length is sensitive to the gas temperature in the cylinder. This, in turn, is affected by heat transfer from the cylinder walls. To avoid uncertainties due to varying cylinder wall temperatures, the engine coolant was preheated to a specific temperature prior to engine start and kept constant during operation by means of a thermostat. The piston was not actively cooled and, to avoid effects of varying piston temperatures, the engine was always run for a specific period before each measurement. These steps were part of the standardized operating procedure developed to provide the same in-cylinder conditions during each measurement. ■

The remaining part of the uncertainty is due to internal noises. These are almost always impossible to avoid. They contribute to the overall variation in the data, which should always be continuously monitored during a measurement campaign. There are two reasons for this. The first is that the variation during the actual measurements gives us a relevant measure of the measurement precision, which will be useful in the error analysis. The second is that this monitoring shows us if the measurement system is stable or not. This aspect was discussed in the measurement system analysis section of Chapter 8, where a control unit broke down halfway through a series of engine measurements. By comparing measurements made at identical conditions before and after this incident it was confirmed that exchanging the control unit did not have a measureable effect on the data. Sometimes equipment fails more slowly, or unknown background factors affect the data in a subtle way. A good general recommendation is therefore to run a *stability check* at regular intervals when collecting data. Depending on the nature of the measurement system, "regular" could mean once a minute, once a day, or even longer intervals.

The stability check consists of repeated measurements at one or several specified conditions. It could be the baseline operating condition of your process, if there is one, or any condition where the system is *expected* to behave repetitively. These measurements are plotted as a function of time in a *control chart*, which is a graphical tool commonly used to monitor the stability of processes. Each time a new stability measurement has been made it is added to the chart.

As shown in Figure 11.7, a control chart contains a centerline that represents the mean value of a measured characteristic. The chart also has two lines representing the upper and lower control limits. These are usually plotted three standard deviations above and below the centerline. Assuming that the error is normally distributed, it is very unlikely to find points outside the control limits. All the points should therefore plot between the lines and the variation should be completely random. If points plot outside the control limits, this is taken as an indication that a background factor influences the result – an influence that must be found and eliminated in order for the system to be considered stable. Note that, since the standard deviation is calculated from all the points in the chart, the control limits will move slightly when a new measurement is added, especially when there are few measurements in the chart.

There are several indications of background effects besides points plotting outside the control limits. We should be watchful of any indication of non-random behavior. For example, if 18 out of 20 points were to plot above the centerline we would be very suspicious that something was wrong, even if all the points were within the control limits. It would suggest that a shift had occurred in the system. Consecutive points that either increase or decrease in magnitude could indicate drift; a periodic pattern is also an indication of

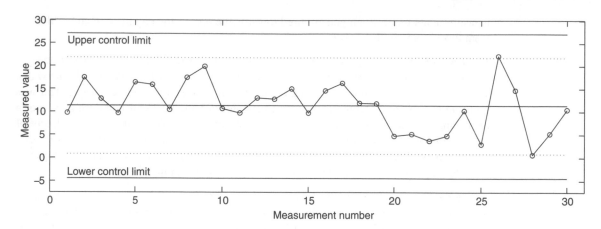

Figure 11.7
Control chart for a stable system. The control limits are three standard deviations from the centerline. The dotted lines are two standard deviations from it.

non-random effects. Montgomery [1] gives the following criteria for detecting background effects in control charts:

- One or more points plot near or outside the control limits.
- At least eight consecutive points either increase or decrease in magnitude.
- At least eight consecutive points plot above or below the centerline.
- Two of three consecutive points plot more than two standard deviations from the control line, but within the control limits.
- Four of five consecutive points plot more than one standard deviation from the control line.
- An unusual or non-random pattern in the data.

> **Exercise 11.4:** Figure 11.8 shows the control chart of a measurement system that, by Montgomery's criteria, has two causes for concern. Identify the issues.

When formulating the data collection plan it is important to consider all parts of the data collection. As mentioned in previous chapters, there is a risk in putting all the efforts into a single experiment. The reason for making an experiment is that we do not know exactly how, or even if, the factors will affect the response. For this reason, the actual experiment is often preceded by a screening phase where the important factors are identified. Experimental results may be so unexpected that the experiment has to be redesigned before continuing. We should, therefore, make a plan where the data collection moves through a number of phases with gradually increasing level of detail. If you are making an experiment based on a factorial design, sequential experimentation (described at the end of Chapter 9) is a good example of this approach. Through the various measurement phases, the plan should specify how the measurements are to be made to mitigate effects of background factors and noises, when and how to run stability checks, and so on. If there are several operators, the plan should also involve practical training to ensure that they collect the data in the same way.

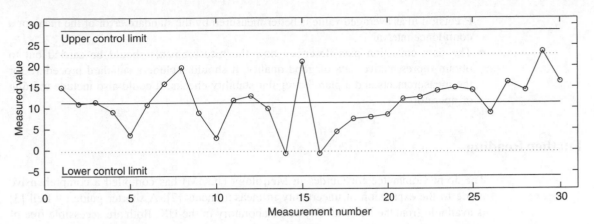

Figure 11.8
Control chart with two causes for concern.

The actual data collection is not particularly challenging. There are, of course, practical problems to be considered but once the data collection plan has been established and everything is working as intended, the measurements generally require little intellectual effort. As always, things can and will go wrong. Having worked through the steps in this chapter, however, we have increased our chances of obtaining data that are useful for our purposes. In the next chapter, we will take a look at the phase where our data are turned from numerical values into carriers of meaning and significance. This involves connecting the measurements back to the research question and to that larger body of knowledge called theory.

11.6 Summary

- To provide explanatory knowledge an experiment should be based on cause variables, not control variables.
- Random error is synonymous with noise and averaging over a sample reduces its influence. External noises arise outside the measurement system and can be handled by isolating the system or reducing the noise (for instance by a strict measurement procedure). Internal noises arise within the measurement system and are more difficult to reduce.
- Bias is synonymous with systematic error and is coupled to the measurement accuracy. It is normally handled through corrections or isolating the system. Randomization is a useful strategy for unknown systematic errors. It is important to consider systematic errors also when using relative data.
- Potential background factors in an experiment may be identified using the Ishikawa diagram (Chapter 10).
- It is useful to apply Ishikawa diagrams, process diagrams and input–output diagrams to the measurement system itself in order to develop strategies for avoiding sources of uncertainty.
- Accuracy is assessed by measuring a known standard. Precision is assessed by replicating measurements. Precision in raw data can be measured by the sample standard deviation.

Precision in an averaged value is better measured by the standard error of the mean or a confidence interval.

- The data collection plan describes how the measurements should be carried out to obtain representative data of good quality. It should include a standard procedure for the measurements and a plan for regular stability checks. It could also include training of operators.

Further Reading

The Joint Committee for Guides in Metrology (JCGM) has compiled a comprehensive guide to the expression of uncertainty in measurements [2]. A shorter guide by Bell [3] is available from the National Physical Laboratory in the UK. Both are accessible free of charge on the internet.

Answers for Exercises

11.1 Potential background variables in the planetarium setup are landmarks such as people or the camera stand. To isolate the effect of the Milky Way other stars in the sky must be removed. You should make sure that there are no sources of polarized light in the planetarium. Note that the diagram does not contain all potential background variables. What would for instance happen if the arena was not level?

11.2 Left for the reader.

11.3 The variation of the length is greater than that of the diameter. Putting the values into Equation 11.7, however, we see that the diameter term of the variance is almost a factor of two greater than the length term. To decrease the total variation we should therefore aim to decrease the standard deviation of the diameter.

11.4 Two of the three observations from 14 to 16 are beyond two standard deviations. The eight points following observation 16 show a trend.

References

1. Montgomery, M.C. (1991) *Introduction to Statistical Quality Control*, 2nd edn, John Wiley & Sons, Inc., New York.
2. BIPM, IEC, IFCC, ILAC, ISO, IUPAC, IUPAP and OIML. (2008) Evaluation of Measurement Data – Guide to the Expression of Uncertainty in Measurement (GUM 1995 With Minor Corrections). JCGM 100:2008, Joint Committee for Guides in Metrology (JCGM), Bureau International des Poids et Mesures (BIPM), Sèvres, France.
3. Bell, S. (2001) A Beginner's Guide to Uncertainty of Measurement. Measurement Good Practice Guide No. 11 (Issue 2). National Physical Laboratory, Teddington, UK.

12 Phase III: Analysis and Synthesis

There are two possible outcomes: if the result confirms the hypothesis, then you've made a measurement. If the result is contrary to the hypothesis, then you've made a discovery.
—Enrico Fermi

Even well planned experiments sometimes produce data that are difficult to interpret. This chapter proposes a procedure for analysis where the data are first processed to obtain meaningful quantities, then arranged into a simple table to make them easy to access and handle, and finally investigated using graphical and mathematical tools. The analysis is discussed relatively briefly, partly because it draws on Chapters 7 and 8 to a great extent, and partly because our efforts in the planning and data collection phases have made the analysis much easier.

The scientific process does not end with data analysis. In many ways, the remaining part is the most important one for the development of science. In the synthesis phase, we relate our findings to the current knowledge and define our contribution to it. This is a central part of publishing our research in journals or at conferences. In a Ph.D. thesis, the synthesis is perhaps the most crucial part. This is where you demonstrate that you have made an original contribution to knowledge and explain what it consists of.

12.1 Turning Data into Information

It is sometimes said that a good experiment analyses itself. It is true that careful planning and data collection make the analysis easier and this is why we devoted so much space to them in the two preceding chapters. But it is equally true that we would not make experiments if we knew the outcome. Effects may turn out to be very weak or there may be unexpected aspects of the data that make them difficult to analyze. And even when the data are good, it may not be straightforward to present the results clearly and concisely.

Even though the preparations in the preceding two chapters will improve our prospects, they do not guarantee success. They say that the road to hell is paved with good intentions. You will know what this means if you spend months planning an experiment, getting your setup to work and collecting data, only to find that the data are insufficient to support a clear conclusion. One way to avoid unpleasant surprises when evaluating the data is to do parts of the analysis already during the data collection phase. This allows us to anticipate potential problems. If the results are not promising, parts of the experiment may have to be

Experiment!: Planning, Implementing and Interpreting, First Edition. Öivind Andersson.
© 2012 John Wiley & Sons, Ltd. Published 2012 by John Wiley & Sons, Ltd.

Figure 12.1
The three phases of research. During the analysis and synthesis phase it may be necessary to revisit the planning phase or the data collection phase in order to build a solid scientific argument.

redesigned. This is, of course, also the reason for the recommendation in the last chapter that we should not collect all the data in one go.

In the last chapter we mentioned that it is sometimes necessary to revisit the planning phase during the data collection phase. In the same way, we may occasionally have to return from the analysis and synthesis phase to the planning or data collection phases. Figure 12.1 shows where in the research process we now find ourselves. The large black arrow points to the planning phase because the knowledge obtained from an experiment that worked well often gives rise to new questions and ideas for research. The smaller arrows indicate that, if the experiment works less well, we may have to reiterate parts of previous research steps.

The analysis begins by looking at the results and trying to find trends and regularities. An initial problem that may have to be overcome is that some experiments do not directly produce numerical data. This is true, for example, when the results come in the form of images. Though effects in such data may be visible in a "soft" way to the eye, they should be boiled down to meaningful numbers to make quantitative analysis easier. As discussed in the last chapter, numbers can be plotted graphically and treated mathematically, whereas soft features only can be qualitatively described.

For this reason, we will break the data analysis down into two activities. The first is data processing, where raw data are turned into a quantitative format. The second activity is the actual analysis, which aims at figuring out what the data can tell us. Processing might involve image analysis or other procedures required to turn soft features in the data into numbers. Once numerical values have been extracted and put into a spreadsheet we can start with the actual analysis, looking for patterns and regularities to arrive at conclusions supported by diagrams and mathematical relationships.

The word analysis has many meanings. In experimental research, it refers to the process of distilling the complex information in raw data into components that carry meaning in the light of our research question. This may be straightforward but in some cases it may require an effort. These components are very similar to the theoretical concepts discussed in Chapter 5. They do not have to be direct properties of the phenomenon under study to be useful. It is sufficient to find a useful representation of the phenomenon – one that can be expressed in numbers. Most of the experiments that have been discussed in this book use such indirect measures to support the conclusions. For instance, in the beetle experiment (Chapter 10, Experiment 1) the phenomenon of interest is the beetle's ability to orient but it is *measured* by the rolling time on the arena.

> **Exercise 12.1:** Recapitulate the example experiments in Chapter 6. For each experiment, identify the phenomenon of interest and the measures by which it is studied. Are any of them direct?

In the ideal case we will already have formed a concrete idea of which measures to use during the planning phase but, sometimes, the analysis takes an unexpected turn. Finding the best way to present your data can require a problem-solving process that is likely to involve a bit of trial-and-error. The following two examples will cast some light over what this process can be like.

Example 12.1: Imagine that you want to investigate the properties of a fuel spray in a spray chamber. You vary certain factors that you suspect will affect the propagation of the spray. The measurements are made using a high-speed video camera, producing a time-resolved sequence of images of the spray during fuel injection. These image sequences contain information about the effects in your experiment but you need to extract the essence of this information and present it in quantitative form. Even if there were space in a journal article to show all the film frames, your readers would not appreciate having to search the information out for themselves. You could start by measuring the distance between the injector and the spray tip in each film frame and plot it as function of time. This will produce a curve describing the spray propagation in each movie. But presenting many such curves together will probably result in a crowded diagram, making it difficult to discern subtle differences between the cases. Since time is not an experimental factor it is probably not the most interesting variable to put on the abscissa of your diagrams. You should instead extract representative numbers from the time trends and plot them, not as a function of time, but as a function of your experimental factors. You could use the maximum spray penetration length, the average spray speed, the deceleration, or other numbers that contain *concentrated* information about the spray propagation. Their dependencies on the experimental factors can be more clearly visualized in diagrams than the direct time trends from the films. ■

Example 12.2: In the diesel engine experiment (Chapter 10, Experiment 2) the data processing consisted in turning information in images into lift-off lengths. Typical images are shown in Figure 10.5. As explained in Example 11.4, the data processing involved minimizing the effects of vibrations. Lift-off lengths were finally extracted from the images by the following algorithm:

Firstly, the images were digitized. Pixels where the signal exceeded the detector noise level were set to white, while the remaining pixels were set to black. A pie-shaped evaluation region was then drawn around each burning spray in the images. The narrow end of the region pointed at the injector and the wide end coincided with the combustion chamber wall. Within each region, the number of white pixels was counted as function of the radial distance from the injector. The lift-off position appeared as an abrupt increase in the white pixel count at a certain distance from the injector. Images often suffer from noise, reflections and other unintended features; this method was developed to work around such problems. It was found to provide consistent results for the entire set of data. ■

In both these examples the data have a natural dimension that is not an experimental factor. In Example 12.1 the dimension is time, since the spray propagation is captured on a film. In the other example the dimension is distance from the injector. When processing the data it is easy to become attached to such natural dimensions and forget that we are interested in the influence of the experimental factors. Coupling our response variables to the experimental factors, or to other variables that are firmly coupled to our hypothesis, makes it easier to demonstrate the relevant patterns in the data.

When the interesting numerical responses have been identified and extracted, the data should be arranged in a table, just as we did in Chapter 9. The rows should correspond to the measurements and each column to a variable. The factor variables should be located in the left-hand part of the table and the response variables to the right. This will make it is easier to toy with the data to look for patterns.

If you are using a Microsoft Excel® spreadsheet you can now select a factor column and sort the values from the smallest to the largest. Make sure that the other columns are sorted according to the chosen column, so the rows remain intact after sorting. This makes it easy to compare the responses to the sorted factor values. Are there any obvious patterns? Does any response seem to correlate positively or negatively with the factor? If the factor is categorical, are there any obvious differences between the groups? Continue like this with all the factor columns. You can also add new columns to the table. For instance, if you expect that there is a quadratic effect, you could add a column containing the square of a factor. The response may also be a function of several factors and you could create columns containing ratios or products of factors to investigate this. Your subject matter knowledge will probably lead you to factor combinations that seem relevant but try to test unconventional ideas too.

At this point you might find that the results are quite different from what you expected. This may force you to follow a new line of thought in your analysis, but it may also mean that you have to take the experiment back to the planning phase and start anew. This is what happened in Experiment 2. The original idea was actually to investigate how the injection pressure affected the lift-off length in an optical engine. When studying high-speed video films of the burning sprays, no distinct effect of the injection pressure could be found. But there was an unexpected feature of the data. As mentioned in Chapter 10, Chartier was puzzled to find that the lift-off length was constantly drifting towards the injector. He wondered if this could be due to the burning sprays replenishing the space between them with hot combustion products and developed Experiment 2 to test this hypothesis. In other words, he took his experiment from the analysis and synthesis phase back to the planning phase, following one of the smaller black arrows in Figure 12.1, before completing the investigation and publishing the results.

Sorting the data and looking for patterns is a useful start but there are, of course, limitations to the relationships you can spot by just looking at a table. This is especially true for large sets of data. The most important step in the analysis is therefore to represent the data graphically.

12.2 Graphical Analysis

It is much easier to see relationships between variables in diagrams than in the numerical values themselves. The most common graphic tools have been introduced in the earlier chapters of this book and are only be briefly recapitulated here. They include tools for

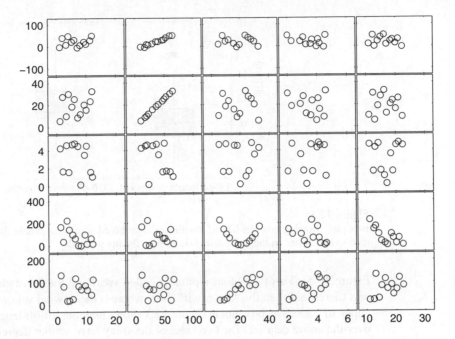

Figure 12.2
Scatter plot matrix connecting five factor variables on the *x*-axis with five response variables on the
y-axis.

investigating distributions, for instance of residuals, but also tools for studying the relationships between two or more variables. When working with continuous variables, the
scatter plot is probably the most popular type of diagram. Depending on the measurement
precision it may or may not be illustrative to connect the points in a scatter plot by lines. If
the data are noisy a regression line may be a better option. In cases when we are interested
in the simultaneous influence of two variables, contour plots are illustrative.

One way to look for patterns and relationships in larger sets of continuous data is to plot
all variables simultaneously in a scatter plot matrix. Figure 12.2 shows an example of such
a diagram, where five factor variables are aligned along the *x*-axis and five responses along
the *y*-axis. This graph provides a convenient overview of all the relationships between the
variables. Most of the panes do not display any particular patterns. In the two rightmost
panes on the top there is no variation in the response at all, except for some noise. It is more
difficult to say if there are any effects in the two panes below them, for instance, because
the noise is so dominant. Four panes in the matrix stand out, showing distinct structures
that indicate relationships. This diagram was made using the PLOTMATRIX function in
MathWorks MATLAB®. Excel has no native functionality for creating this type of diagram.

For categorical variables, bar plots are more useful than scatter plots. An example of a
bar plot is found in the widowbird example of Chapter 6, where Figure 6.12 shows the
birds' mating success before and after tail treatment. If we want to compare both the central
tendency and the distribution of two or more data sets, the box plot is a more useful graph.

Example 12.3: In Experiment 2 (Chapter 10), the lift-off lengths tended to be shorter
on one side of the burning sprays. The tendency seemed to correlate with the motion of
the in-cylinder air, which was rotating about the cylinder centerline (counter-clockwise in

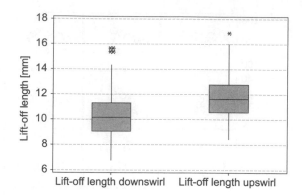

Figure 12.3
Box plot showing how the lift-off lengths are distributed across the whole data set. Lift-off lengths tend to be shorter on the downswirl side than on the upswirl side.

Figure 10.5). This type of air motion is called swirl and the downwind side of the spray was therefore called the "downswirl' side, whereas the upwind side was referred to as the "upswirl' side. The box plot in Figure 12.3 shows how the lift-off lengths were distributed over the entire data set. The two sides of the spray have similar degrees of dispersion and skew but the median value on the downswirl side is about 15% lower than that on the upswirl side. A similar trend had previously been reported from another study [1]. Interestingly, Experiment 2 used a stronger swirling flow and also showed a greater difference in lift-off length between the two sides of the spray. These results seemed to support the hypothesis that hot burned gases affected the lift-off length. This is because the airflow was expected to displace hot products from the upswirl to the downswirl side of the spray. Under the research hypothesis, it is fair to assume that a greater quantity of hot gases on the downswirl side of the spray will shorten the lift-off length there. ∎

You may have noted that the responses in Figure 12.3 are not plotted as a function of the experimental factors. As explained in the example, the plot still provides support for the scientific argument that hot gases affect the lift off length. This is because the variables on the abscissa are firmly coupled to the research hypothesis. Interestingly, the idea that the lift-off length could be asymmetric about the spray did not occur when designing the experiment. The effect was discovered when processing the data and was included in the analysis since it provided additional support for the research hypothesis. It should also be noted that the variable on the abscissa is categorical, although the experimental design was based on continuous numerical factors. This illustrates that there is nothing automatic about data analysis. Even data from carefully designed experiments often contain unexpected features.

The final aim of the analysis is to build a scientific argument, where your data is the evidence that demonstrates the soundness of your conclusions. Clear diagrams are probably the most intuitive and convincing way to support your conclusions; this is the reason why the discussion in almost all research papers is built around them. But a diagram is not an argument in its own right. It only describes what happened in your experiment. Looking at our example experiments we realize that trends and differences in the data make sense only in the light of the research hypothesis.

So, although the analysis phase in many ways is a process of finding the diagrams that most clearly display the essential features of your data, this process must be firmly guided by your research question. Did the results confirm or disprove your initial hypothesis? Can established theories explain the results, or are there discrepancies? Do you need further hypotheses and experiments to confirm your findings? Your diagrams must answer questions like these if you are to build a coherent scientific argument from them.

12.3 Mathematical Analysis

To further investigate the nature of the relationships in your data you may continue with more quantitative, mathematical techniques, such as the ones presented in Chapter 8. When working with categorical data, the classical significance tests are often useful for demonstrating effects in a more unambiguous way than diagrams do. In this book we have only utilized the most fundamental tests. If these are not sufficient for your needs, the statistical literature contains a wealth of significance tests for various types of data.

For continuous numerical data, correlation and regression analysis are useful techniques for quantifying relationships between variables. Let us look at an example of how regression analysis was applied to strengthen the conclusions from Experiment 2.

Example 12.4: As explained in Chapter 10, the measured lift-off lengths in Experiment 2 were compared with an empirical formula, based on an extensive database from a spray chamber. According to this formula, the temperature, T, of the air surrounding the spray had a dominant effect on the lift-off length. More specifically, the lift-off length was proportional to $T^{-3.74}$. In Experiment 2, the lift-off length shortened with increasing air temperature, but less than the empirical formula predicted. Regression analysis yielded a temperature exponent -1.55, substantially lower than the value of -3.74. The temperature dependence of the lift-off length was clearly weaker in the optical engine than in the spray chamber, indicating that the processes differed between these environments. To further support this conclusion, the observed lift-off values were plotted against those predicted by the empirical formula. The diagram is shown in Figure 12.4. The engine data obviously do not agree very well with the formula. ∎

In this example, regression analysis was used to quantify the effect of a variable, just as we did when analyzing response surface experiments in Chapter 9. When compared to simply plotting the data, this mathematical analysis allows for a more unambiguous comparison of results between different experiments.

Before leaving mathematical analysis it could be useful to briefly mention *multivariate data analysis*. Data from experiments tend to be structured because experiments are designed to expose how one set of variables responds to structured changes in another. In some situations, however, we are faced with unstructured data. This situation could arise if it is impossible to control certain experimental variables independently of others. It is also common to search for relationships in large sets of historical data that are not the result of structured experiments. This could be a useful preparation for an experiment, as patterns in such data sets may help us to generate new hypotheses. (Remember that the research process is circular: analysis can serve as a preparation for the planning phase of an upcoming experiment.) Whenever we encounter unstructured data, multivariate techniques may be useful.

Multivariate techniques are statistical methods for investigating large sets of variables that are related to each other in ways that make it impossible to interpret their effects separately. A common approach is to look for a smaller set of underlying variables that

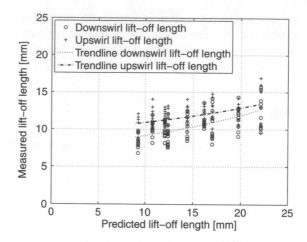

Figure 12.4
Measured lift-off lengths plotted versus the ones predicted by an empirical formula. If the formula
had applied to the measurements the points would have fallen on a line with a slope of one.

summarizes the original ones. It is often possible to express most of the variation in the data
set using a few orthogonal "dummy variables" that make it much easier to see relationships
in the data. Unfortunately, a detailed discussion of multivariate techniques falls outside the
scope of this book, but it is useful to know that they exist. If you think that they could
be useful in your research, there are many textbooks on the subject to choose from. Hair,
Anderson, Tatham, and Black [2] is one example that provides a useful overview.

12.4   Writing a Scientific Paper

The last step of the analysis phase is to write a scientific paper. This may seem like a
separate activity but you will soon realize that explaining your results to others is essential
for your own understanding. The most important thing to make clear is that a scientific
paper is not a laboratory report – it is a scientific argument. You are not writing a summary
of activities carried out in a laboratory to demonstrate that you master a technique or a
piece of theory. You are not expected to give a full account of everything that you did in the
laboratory – mistakes, detours, troubleshooting, and so on – unless they have bearing on
the conclusions and contribute to formulating a coherent scientific argument. The readers
will not be interested in how difficult it was for you to arrive at your conclusions. They read
to take part of the conclusions and want to see the evidence for them.

Writing is part of the analysis phase because it is a way to explain to yourself, as well
as others, how your conclusions are connected with your data. If you cannot explain this,
the analysis is not complete. Writing also involves putting your findings in relation to the
knowledge in your field. As a result of your experiment this knowledge will have changed,
if only a little, and your paper should make this change clear. Writing thereby also marks the
beginning of the synthesis phase: that of adding a specific piece of research to the general
scientific knowledge. These two phases are treated as one in this chapter as they are two
sides of a coin: your research question addressed a limitation in the current knowledge, and
the answer thereby makes a natural contribution to that knowledge.

The writing process begins with formulating the argument that you are going to make and does not end until the paper has been published. Your manuscript will go through a series of reviews and revisions along the way. Before submitting it to a journal or a conference you will probably discuss it with your colleagues and your academic supervisor. After submission, fellow scientists in your research community will review your text. These are appointed by the editor of the journal and will remain anonymous to you. Their comments will highlight issues with anything from the research question to the presentation of the results. The aim is to ensure that you have made an original contribution and that your paper is up to the accepted quality standards of your field.

We are not going to discuss all technical aspects of writing here, as there are whole books on this subject. There are also plenty of guides to technical writing on the Internet, as well as journal-specific author guidelines provided at journal home pages. Though there are minor differences between publishers, for example in reference style or in allowing authors to write in the first person, research papers generally adhere to a standardized format that is relatively fixed across disciplines. We will briefly discuss the parts of a scientific paper, as it is important to understand how results are communicated within the scientific community. A paper is generally built around the following rough headings:

- Background
- Method
- Results
- Analysis
- Conclusions.

The purpose of the *background* section is to present your research question. It explains the central problem of your study and justifies why it is worthy of closer investigation. It also describes the scope of your study. Which parts of the problem did you address and which parts did you leave out? The reader must be able to understand from your description what the research question is and why you have chosen to answer it. This naturally requires some discussion of the current status of knowledge with references to the relevant literature. It is good practice to start the background section with a general picture of the problem area and then, over a couple of paragraphs, gradually move closer to the specific problem that you have studied. This sets the scene for your research and guides your readers from the common ground that you share with them to the point that is of specific interest in your paper.

The background section also contains theory. Although scientific papers are written for specialists, it may be necessary to explain certain ideas and concepts that were used in the design or the analysis of your experiment. Any definitions that are needed to understand the line of reasoning should also be presented here. If the theory requires significant space, it may be appropriate to divide the background section into an introduction and a theory part.

The purpose of the *method* section is to explain your practical approach to the research problem. Here, you describe which data you collected, how you collected them and how you arrived at the results. It covers instruments and measurement techniques, experimental setups, materials, measurement procedures, as well as how you processed the data before the analysis. You must explain these things at a level of detail that allows the reader to replicate your study and arrive at similar results. The method section should also describe and explain the experimental design. This will include a discussion of how you handled background factors and noises.

The *results* section contains all the evidence that your conclusions are based on. It is not acceptable to present an arbitrary collection of tables and diagrams. The data must be presented in terms of the research problem. This will make it possible for the reader to follow your line of reasoning through the paper. When you use statistics, you should explain the use of the specific statistical techniques. It must be clear to the reader that they have been used correctly. Make sure that you include all the data that bear on your conclusions. If measurements are to be left out there must be very good reasons for it; these must be clearly stated. Recall the discussion about Robert Millikan in Example 6.5, who was accused of academic dishonesty for leaving out observations that he deemed unreliable. Your paper should give an honest and objective account of the support for your conclusions.

The *analysis* section interprets the data for the reader and explains how they answer the research question. The conclusions must be entirely supported by the data presented in the results section. If you want to discuss ideas that are not a direct consequence of the data, present them as new hypotheses and make clear that they require further investigation. This section should state whether the data support the initial hypothesis or not, explain the implications of the results, and point to further studies that are needed to provide a more complete answer to the research question. If the paper treats several subproblems, it may be perceived as repetitive to discuss the implications of the results in a separate section. For this reason, the two sections are sometimes merged into a combined results and analysis section.

The *conclusions* section simply contains a condensed summary of what has been learned from your study.

These sections are preceded by the paper title, a list of authors, and an abstract. The title should convey the essence of the paper in a single phrase. It may be a simple statement of the central conclusion or the research question. The title may also refer to methods or techniques used in the paper, if this is relevant to the reader. It is important to understand that researchers search the literature in bibliographic databases. This is where your paper is most likely to be accessed after publication. A literature search is based on keywords that are matched with specific parts of papers. You can search in the titles, the abstracts, the author lists, and so on. The search engine usually lists the titles of the papers that match the keywords. This is why it is important to choose a title that accurately targets a relevant part of the research community. They must understand from the title if your paper is worth a closer look.

The first step in taking a closer look is to read the abstract, which is usually a single paragraph summary of the paper. It describes the central research question, the method used to address it, and the central conclusions. If the abstract confirms the impression that the paper is relevant, the person making the literature search will download the actual paper from the publisher and read it in its entirety. Once you have made a literature search yourself, you will realize how important it is to convey the central message of the paper in the title and abstract.

At the end of the paper there should be a list of research papers that you have referred to in the text. The specific format of this reference list varies between journals and is therefore described at each publisher's home page. There is usually an acknowledgement section too, where it is customary to thank relevant funding bodies for research grants. Such organizations often search the databases to evaluate the research that they have supported. For this reason, they may demand that researchers state the source of their funding when publishing their work.

Keep in mind that a research paper does not aspire to be a work of fine literature. It is supposed to be a concise report written in factual prose. Above all, it should be clear and brief. A long manuscript may even be reason for refusal. The best way to understand which style is appropriate for a given journal is to read a few of its articles. One of the main challenges is often to maintain a clear line of argument all the way from the background through to the conclusions. There are probably as many strategies to achieve this as there are authors, but you may find the following technique useful.

Before you start to write you need to make clear to yourself exactly what your scientific argument is. The central message of your paper is, of course, most clearly conveyed by your conclusions. For this reason it could be useful to start by writing down the conclusions and then finish the remaining sections in the reverse order to that in which they will appear in the paper. This is because each section provides the necessary support for the subsequent section: the analysis leads to the conclusions, the data is the basis of the analysis, the methods provide the data, and the research question determines which methods to use. This method of reverse writing allows you to follow the river back to its source, so to speak. If you begin by writing the background section, there is a greater risk that your line of argument becomes diluted in detours and digressions before you reach the end. There is also a risk that you become tired before you reach the sections that are most central to your argument. The analysis and conclusions sections are where you need to be the most sharp and focused in your writing. As these sections are so important, there are even people who habitually read research papers in the reverse order, starting with the conclusions and finishing with the background.

Practically, when writing a paper in this way, you begin by printing out the diagrams that demonstrate the validity of your conclusions. Pin them to the wall by your desk so that they are constantly available to you during the writing process. Create a text document and put all the section headings down in the right order. Write your conclusions under the final heading, briefly and to the point, in the form of a bulleted list. Copy this list and paste it under the analysis heading. Under each bullet, write down the main points of the analysis that lead to the conclusion. Do not worry about language and style at this point; just concentrate on putting down all the relevant information. Then move to the results section and describe the data and diagrams that underlie the analysis under each point. Continue backwards through the headings like this, describing the methods used to collect the data in the methods section and how the research problem led to these methods under the background heading. Whatever approach you take to writing, do not begin to write before you have all the conclusions in place. Writing a paper at the same time as you process and review the data is a safe method for obtaining a vague and confusing manuscript.

Another useful technique is to present your work orally to colleagues before starting to write. This presentation should include all the parts of a scientific paper in the right order. The feedback you receive will reveal whether you manage to convey your argument or not. Once you have worked out a clear and concise presentation, what remains is basically to transfer it into a text document and put on a finishing touch.

Authoring a research paper can be compared to being a tour guide. Needless to say, your readers cannot read your thoughts, so you must explain your thinking to them. Although the contents of the paper are summarized under standard headings, you must tie these together into a comprehensible chain of thoughts. After introducing the research idea, you must explain how it led to your approach, how the approach resulted in the data, and how your analysis led you to the conclusions. Reading a paper where the author has neglected to guide the reader is like walking through a museum where historical objects are on display without descriptions of what they are, where they came from, and what we might learn from them.

Finally, the peer review process is an integral and important part of writing up your research. It can be frustrating to obtain anonymous, critical comments on a text that is the result of great personal effort. Review comments are often brief and may sometimes be perceived as impolite, especially before you are used to the process. It is important not to take the criticism personally. Reviewing a paper is time consuming and, since referees are fellow researchers with many other tasks on their hands, it is understandable if they do not have time to formulate their comments as diplomatically as might be wished. We should rather see each comment as a valuable opportunity to improve the quality of our paper. They often indicate points where we have failed to explain our thoughts clearly, or where important information is missing. Intense work with a manuscript tends to make us blind to certain aspects of the text and it is valuable to get comments from people who have read it with fresh eyes.

A manuscript will typically be reviewed by several anonymous referees. After obtaining the review results you are expected to do two things. Firstly, you should write a rebuttal where you give detailed responses to each of the comments. Secondly, you should make appropriate updates to your manuscript. The rebuttal should indicate where these changes have been made, and it is good practice to highlight the changes in the manuscript so that the referees may find them easily. After submitting the revised manuscript and the rebuttal to the publisher, there is generally another period of waiting before you are notified of the journal's decision to publish or not to publish your paper. The publishing process usually takes additional time, so it may be a year between the original submission and the publication of your article in the journal.

In the mean time you will of course be fully occupied with your next research task. In all likelihood you entered the planning phase of this task when analyzing the data from the previous one, because analysis has a tendency to uncover new aspects of a problem and generate new ideas for experiments. As illustrated by Figure 12.1, the circle of research is never completed. In fact, it is impossible to say where it begins and ends.

Communicating one's results within the scientific community is one of the most important parts of doing research. Despite this, Ph.D. students in some places are not required to publish their work. Since writing is as difficult a skill to learn as anything else in research, it is doubtful whether a person can be considered to be an accomplished researcher before being published. Imagine calling yourself a journalist if you had never written a newspaper article. As research students are in training to become autonomous researchers, my recommendation is to publish as much as possible of your work prior to graduating. Writing papers is a central research skill and, as any other skill, it can only by learned by performing it. An additional advantage is that published papers confirm that your work meets the accepted quality standards in your field. When defending your thesis one day, your peer reviewed journal articles actually constitute objective proof of your research skills that can be used to meet the potentially more subjective opinions of the members of your grading committee.

12.5   Writing a Ph.D. Thesis

The Ph.D. degree is awarded for an original contribution to knowledge. One important function of the Ph.D. thesis is, therefore, to demonstrate that you have made such a contribution. This is where you put the parts of your work together to formulate higher-level conclusions. These are then related to the work of others to define, in a wider sense, how your research has developed the knowledge in your field. A thesis thus contains a more general synthesis than that in a research paper.

As the thesis is the last stop on your journey towards the Ph.D. it is appropriate to briefly discuss its parts here. Though these parts are similar to those of a scientific paper, their functions are different. The following lies close to Phillips and Pugh [3]. Originally, a master's degree was a license to practice, whereas a doctor's degree was a license to teach at university level. To teach with authority, faculty members obviously need to have a deep knowledge of their subject. They need to have a grasp of the current knowledge and be capable of extending it. You demonstrate these abilities through the parts of your thesis.

Just as a scientific paper, a Ph.D. thesis begins with a *background*. The main purpose, however, is not to prepare the reader for the upcoming analysis. It is to demonstrate that *you* master the theory in your field to the full professional standard. This is normally accomplished by a literature review of your field but it is not sufficient to just list relevant studies one after the other. To show that you are in command of your subject you must organize the knowledge in a coherent and interesting way. It is important to demonstrate that you understand central theoretical concepts, that you can identify research trends and point to areas of knowledge where there is room for improvement. You should also show that you are able to evaluate the contributions of others and justify your criticism. In addition to this, the background chapter should motivate your own research by spelling out in detail exactly what you are studying and why.

The *method* chapter of a thesis contains similar information to that in a research paper but its function is mainly to show that *you* understand and master all the practical aspects of experimentation. For example, in addition to describing which measuring instruments you used, you must devote space to explaining their working principles. Instead of just describing your analysis techniques, you explain the fundamental principles behind them, including a critical discussion of their limitations.

Just as in a paper, *results* and *analysis* chapters follow the background and method parts. The extent of these will vary depending on the format of the thesis. If you are writing a monograph, all the data and analyses must be presented in depth within the thesis. In some places, it is more common to write a summary thesis with the published papers as appendices. In such theses the results and analysis chapters are considerably shorter and contain ample references to the papers instead.

The most important part of your thesis is the *contribution* chapter. This is where you clearly spell out how you have changed the knowledge in your field. In this chapter, you put the conclusions from your work into a wider context. You also describe the limitations of your research, with appropriate suggestions for new work. One might ask why Ph.D. students are expected to evaluate their own work. Isn't this the grading committee's job? Well, the thesis serves to demonstrate that you have become a fully professional researcher with a firm grasp of your field. As a part of this you must show that you are capable of evaluating the impact of research, your own as well as that of others.

The contribution defines the synthesis between the parts of your research and the greater body of knowledge in your field. To put it simply, it explains why and in what way the contents of the background chapter have now changed as a result of your work. From now on, your successors will face a different situation because they have to take account of your research when they plan their own.

12.6   Farewell

We have now reached the end, not only of the research process but also of this book. I hope that it has conveyed ideas and perspectives that will not only make your research easier but also enhance your experience of doing research. Whatever path you choose to take in the future, I hope that some bits of this book will remain helpful when approaching new challenges.

Personal development consists of two parts that mutually feed each other. The first is the development of skills. Skills can be seen as a toolbox that comes about through practice and reflection. It is my hope that this book has contributed to developing some of your skills. The other part is the development of new ideas. This book cannot develop new ideas for you, nor can any other. You must find your own ways to let them grow from within. It is important to realize that ideas are the material on which we apply our skills. Without them our skills are useless. Ideas are the very stuff from which we build the future.

I hope that you will find ample opportunity to apply and develop your skills. And do not let the occasional chance to develop an original idea pass you by. As Douglas Adams put it, "You live and learn. At any rate, you live".

12.7   Summary

- Analysis can be seen as distilling the complex information in raw data into components that are meaningful in light of your research question. It commences with data processing and finishes with the actual analysis.
- Processing includes any steps necessary to turn the raw data into a quantitative format.
- The measures used in the analysis may be an indirect representation of the phenomenon under study. They are thereby similar to the theoretical concepts discussed in Chapter 5.
- We should connect our response variables to the experimental factors or other variables that are firmly coupled to the research question.
- It is good practice to begin the analysis by arranging the processed data in a table and look for patterns. The hypothesis will make you expect certain patterns but it does no harm to take a wider view. This stage should be followed by graphical and mathematical analysis.
- Multivariate techniques can be used to analyze non-structured data.
- A research paper presents a scientific argument. It begins by clearly stating a research question and ends with conclusions that answer it. It also describes the methods and gives all the data that underlie the conclusions.
- The parts of a Ph.D. thesis are similar to those of a paper but they focus more on demonstrating the research skills of the student.
- The synthesis consists of specifying your scientific contribution: putting the parts of your research together and relating them to the larger body of knowledge in your field. It is a central element of research papers and theses.

Further Reading

Additional analysis techniques can be found in most statistics books. Box, Hunter and Hunter [4] focus on techniques in an experimental context. Hair, Anderson, Tatham, and

Black [2] introduce a range of multivariate techniques. For further information on the Ph.D. process, the reader is recommended to read Phillips and Pugh [3].

Answers for Exercises

12.1 Two of them use direct measures. In Example 6.2, Feynman directly studies the effect of cold on the resilience of the O-ring. In Example 6.3, Mendel studies inheritance by direct studies of characteristics. The other examples use indirect measures. 6.1: Harvey studied the motion of the blood, for instance by looking at the color of the heart when obstructing the flow of blood to or from it. 6.4: Hershey and Chase drew conclusions about the genetic material of viruses by seeing where a radioactively labeled material ended up after infection. 6.5: Millikan studied the "atomic" nature of the electric charge by making charged drops hover in an electric field. 6.6: Chartier studied the impact of overleaning on the emissions of hydrocarbons by looking at the reactivity and composition of the fuel–air mixture near the injector. 6.7: Andersson studied the evolution of male ornaments by sexual selection through counting the number of active nests in the territories of tail-treated widowbirds.

References

1. Musculus, M.P.B. (2003) Effects of the In-Cylinder Environment on Diffusion Flame Lift-Off in a DI Diesel Engine. SAE Technical Paper, 2003-01-0074. doi: 10.4271/2003-01-0074.
2. Hair, J.F., Anderson, R.E., Tatham, R.L., and Black, W.C. (1998) *Multivariate Data Analysis*, 5th edn, Prentice-Hall, Upper Saddle River (NJ).
3. Phillips, E.M. and Pugh, D.S. (1994) *How to Get A PhD: A Handbook For Students and Their Supervisors*, 2nd edn, Open University Press, Buckingham.
4. Box, G.E.P., Hunter, J.S., and Hunter, W.G. (2005) *Statistics for Experimenters: Design, Innovation, and Discovery*, 2nd edn, John Wiley & Sons, Inc., Hoboken (NJ).

Appendix

Standard Normal Probabilities

Add bold values in the first column and row to obtain Z-values with two decimal places. The table gives the probability of drawing a value less than Z.

Z	0.00	0.01	0.02	0.03	0.04	0.05	0.06	0.07	0.08	0.09
0.0	0.5000	0.5040	0.5080	0.5120	0.5160	0.5199	0.5239	0.5279	0.5319	0.5359
0.1	0.5398	0.5438	0.5478	0.5517	0.5557	0.5596	0.5636	0.5675	0.5714	0.5753
0.2	0.5793	0.5832	0.5871	0.5910	0.5948	0.5987	0.6026	0.6064	0.6103	0.6141
0.3	0.6179	0.6217	0.6255	0.6293	0.6331	0.6368	0.6406	0.6443	0.6480	0.6517
0.4	0.6554	0.6591	0.6628	0.6664	0.6700	0.6736	0.6772	0.6808	0.6844	0.6879
0.5	0.6915	0.6950	0.6985	0.7019	0.7054	0.7088	0.7123	0.7157	0.7190	0.7224
0.6	0.7257	0.7291	0.7324	0.7357	0.7389	0.7422	0.7454	0.7486	0.7517	0.7549
0.7	0.7580	0.7611	0.7642	0.7673	0.7704	0.7734	0.7764	0.7794	0.7823	0.7852
0.8	0.7881	0.7910	0.7939	0.7967	0.7995	0.8023	0.8051	0.8078	0.8106	0.8133
0.9	0.8159	0.8186	0.8212	0.8238	0.8264	0.8289	0.8315	0.8340	0.8365	0.8389
1.0	0.8413	0.8438	0.8461	0.8485	0.8508	0.8531	0.8554	0.8577	0.8599	0.8621
1.1	0.8643	0.8665	0.8686	0.8708	0.8729	0.8749	0.8770	0.8790	0.8810	0.8830
1.2	0.8849	0.8869	0.8888	0.8907	0.8925	0.8944	0.8962	0.8980	0.8997	0.9015
1.3	0.9032	0.9049	0.9066	0.9082	0.9099	0.9115	0.9131	0.9147	0.9162	0.9177
1.4	0.9192	0.9207	0.9222	0.9236	0.9251	0.9265	0.9279	0.9292	0.9306	0.9319
1.5	0.9332	0.9345	0.9357	0.9370	0.9382	0.9394	0.9406	0.9418	0.9429	0.9441
1.6	0.9452	0.9463	0.9474	0.9484	0.9495	0.9505	0.9515	0.9525	0.9535	0.9545
1.7	0.9554	0.9564	0.9573	0.9582	0.9591	0.9599	0.9608	0.9616	0.9625	0.9633
1.8	0.9641	0.9649	0.9656	0.9664	0.9671	0.9678	0.9686	0.9693	0.9699	0.9706
1.9	0.9713	0.9719	0.9726	0.9732	0.9738	0.9744	0.9750	0.9756	0.9761	0.9767
2.0	0.9772	0.9778	0.9783	0.9788	0.9793	0.9798	0.9803	0.9808	0.9812	0.9817
2.1	0.9821	0.9826	0.9830	0.9834	0.9838	0.9842	0.9846	0.9850	0.9854	0.9857
2.2	0.9861	0.9864	0.9868	0.9871	0.9875	0.9878	0.9881	0.9884	0.9887	0.9890
2.3	0.9893	0.9896	0.9898	0.9901	0.9904	0.9906	0.9909	0.9911	0.9913	0.9916
2.4	0.9918	0.9920	0.9922	0.9925	0.9927	0.9929	0.9931	0.9932	0.9934	0.9936
2.5	0.9938	0.9940	0.9941	0.9943	0.9945	0.9946	0.9948	0.9949	0.9951	0.9952
2.6	0.9953	0.9955	0.9956	0.9957	0.9959	0.9960	0.9961	0.9962	0.9963	0.9964
2.7	0.9965	0.9966	0.9967	0.9968	0.9969	0.9970	0.9971	0.9972	0.9973	0.9974
2.8	0.9974	0.9975	0.9976	0.9977	0.9977	0.9978	0.9979	0.9979	0.9980	0.9981
2.9	0.9981	0.9982	0.9982	0.9983	0.9984	0.9984	0.9985	0.9985	0.9986	0.9986
3.0	0.9987	0.9987	0.9987	0.9988	0.9988	0.9989	0.9989	0.9989	0.9990	0.9990
3.1	0.9990	0.9991	0.9991	0.9991	0.9992	0.9992	0.9992	0.9992	0.9993	0.9993
3.2	0.9993	0.9993	0.9994	0.9994	0.9994	0.9994	0.9994	0.9995	0.9995	0.9995
3.3	0.9995	0.9995	0.9995	0.9996	0.9996	0.9996	0.9996	0.9996	0.9996	0.9997
3.4	0.9997	0.9997	0.9997	0.9997	0.9997	0.9997	0.9997	0.9997	0.9997	0.9998
3.5	0.9998	0.9998	0.9998	0.9998	0.9998	0.9998	0.9998	0.9998	0.9998	0.9998

Experiment!: Planning, Implementing and Interpreting, First Edition. Öivind Andersson.
© 2012 John Wiley & Sons, Ltd. Published 2012 by John Wiley & Sons, Ltd.

Z	0.00	0.01	0.02	0.03	0.04	0.05	0.06	0.07	0.08	0.09
3.6	0.9998	0.9998	0.9999	0.9999	0.9999	0.9999	0.9999	0.9999	0.9999	0.9999
3.7	0.9999	0.9999	0.9999	0.9999	0.9999	0.9999	0.9999	0.9999	0.9999	0.9999
3.8	0.9999	0.9999	0.9999	0.9999	0.9999	0.9999	0.9999	0.9999	0.9999	0.9999
3.9	1.0000	1.0000	1.0000	1.0000	1.0000	1.0000	1.0000	1.0000	1.0000	1.0000

Probability Points for the *t*-Distribution

The table assumes a one-sided test

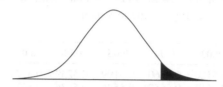

	Tail probability									
ν	0.4	0.25	0.1	0.05	0.025	0.01	0.005	0.0025	0.001	0.0005
1	0.325	1.000	3.078	6.314	12.706	31.821	63.657	127.32	318.31	636.62
2	0.289	0.816	1.886	2.920	4.303	6.965	9.925	14.089	22.327	31.599
3	0.277	0.765	1.638	2.353	3.182	4.541	5.841	7.453	10.215	12.924
4	0.271	0.741	1.533	2.132	2.776	3.747	4.604	5.598	7.173	8.610
5	0.267	0.727	1.476	2.015	2.571	3.365	4.032	4.773	5.893	6.869
6	0.265	0.718	1.440	1.943	2.447	3.143	3.707	4.317	5.208	5.959
7	0.263	0.711	1.415	1.895	2.365	2.998	3.499	4.029	4.785	5.408
8	0.262	0.706	1.397	1.860	2.306	2.896	3.355	3.833	4.501	5.041
9	0.261	0.703	1.383	1.833	2.262	2.821	3.250	3.690	4.297	4.781
10	0.260	0.700	1.372	1.812	2.228	2.764	3.169	3.581	4.144	4.587
11	0.260	0.697	1.363	1.796	2.201	2.718	3.106	3.497	4.025	4.437
12	0.259	0.695	1.356	1.782	2.179	2.681	3.055	3.428	3.930	4.318
13	0.259	0.694	1.350	1.771	2.160	2.650	3.012	3.372	3.852	4.221
14	0.258	0.692	1.345	1.761	2.145	2.624	2.977	3.326	3.787	4.140
15	0.258	0.691	1.341	1.753	2.131	2.602	2.947	3.286	3.733	4.073
16	0.258	0.690	1.337	1.746	2.120	2.583	2.921	3.252	3.686	4.015
17	0.257	0.689	1.333	1.740	2.110	2.567	2.898	3.222	3.646	3.965
18	0.257	0.688	1.330	1.734	2.101	2.552	2.878	3.197	3.610	3.922
19	0.257	0.688	1.328	1.729	2.093	2.539	2.861	3.174	3.579	3.883
20	0.257	0.687	1.325	1.725	2.086	2.528	2.845	3.153	3.552	3.850
21	0.257	0.686	1.323	1.721	2.080	2.518	2.831	3.135	3.527	3.819
22	0.256	0.686	1.321	1.717	2.074	2.508	2.819	3.119	3.505	3.792
23	0.256	0.685	1.319	1.714	2.069	2.500	2.807	3.104	3.485	3.768
24	0.256	0.685	1.318	1.711	2.064	2.492	2.797	3.091	3.467	3.745
25	0.256	0.684	1.316	1.708	2.060	2.485	2.787	3.078	3.450	3.725
26	0.256	0.684	1.315	1.706	2.056	2.479	2.779	3.067	3.435	3.707
27	0.256	0.684	1.314	1.703	2.052	2.473	2.771	3.057	3.421	3.690
28	0.256	0.683	1.313	1.701	2.048	2.467	2.763	3.047	3.408	3.674
29	0.256	0.683	1.311	1.699	2.045	2.462	2.756	3.038	3.396	3.659
30	0.256	0.683	1.310	1.697	2.042	2.457	2.750	3.030	3.385	3.646
40	0.255	0.681	1.303	1.684	2.021	2.423	2.704	2.971	3.307	3.551
70	0.254	0.678	1.294	1.667	1.994	2.381	2.648	2.899	3.211	3.435
130	0.254	0.676	1.288	1.657	1.978	2.355	2.614	2.856	3.154	3.367
∞	0.253	0.674	1.282	1.645	1.960	2.326	2.576	2.807	3.090	3.291

Index

χ^2-distribution 164

Abdera 37
acceleration 62
acceleration due to gravity 58
accuracy 116, 159, 171, 236, 238, 245–6
ad hoc modification 21, 22
Aegean Sea 34–8
Alexander the Great 40
aliasing 187
analysis of variance (ANOVA) 155–9, 161–2,
    165, 175, 176, 246
Anaximander 37
Anaximenes 37
Andersson, Malte 106–7
Aristarchus 38
Aristotle 14, 21, 27, 40, 41, 45, 54, 55, 62, 88,
    227
art of discovery 5–7
axioms 23–4

background factor 109, 175, 176, 185, 206, 230,
    242, 248, 249, 250, 261
back-of-an-envelope calculation 227
bacteriophage 96–8
Bartholin, Erasmus 26
bias 237
black box testing 88
blinded study 176
block variable 185
blood circulation 88–90
Box, George 186
Box–Behnken design 205–6
Brahe, Tycho 44–5
Brewster, David 26–7, 88

categorical factors 165, 176–85, 199
causation 85–7
cause-and-effect table 224–5
Cavendish experiment 84–5
center point 199
centering 116
central composite design 199
central limit effect 126–8, 245
central limit theorem 128, 134, 149, 162, 247

central tendency 117
Challenger Space Shuttle 90–2
Chartier, Clément 102, 216–17
Chase, Martha 96–8
checklists 229–31
chlorophyll 4
coded values 184
collinearity 181
combustion 27–8, 72–4
comparing two samples 150–4
confidence intervals 132–4
confidence level 149
confounding 187, 188
    determining pattern 188–90
continental drift 85
continuous factors 195–207
control chart 249–50
control group 22, 87, 106–7, 108–9, 150, 176
controlled experiments 17, 108, 176, 177,
    230
Copernicus, Nicolaus 20, 37–8, 41–2
Copernican system 20, 21, 41–2, 74
correlation 85–7, 165–6, 259
creativity 65, 69, 77–9, 223
crossover experiment 177
cumulative distribution function 123
cumulative probability 130

da Vinci, Leonardo 78
Darwin, Charles 28, 85, 105
data analysis 113–15, 253
    graphical analysis 256–9
    information from data 253–6
    mathematical analysis 259–60
data collection 13, 231, 235
    measurement system analysis 244–8
    measurement system development 238–44
    measurement uncertainty 236–8
    planning 248–51
    understanding 235–6
data mining 87
data presentation 255
De Nova Stella 44
deduction 14–15
defining relation 188–9

Experiment!: Planning, Implementing and Interpreting, First Edition. Öivind Andersson.
© 2012 John Wiley & Sons, Ltd. Published 2012 by John Wiley & Sons, Ltd.

degrees of freedom 120, 135, 144, 157, 163–4, 185, 194, 199, 200
Democritus 37, 45, 49, 98
density function 123
dependent samples 150, 154
descriptive statistics 116–22
designing experiments *see* experimental design
Dialogues Concerning Two New Sciences 54
diesel engines 72–4, 75, 101–5, 159, 200–6, 216–18
discovery, art of 5–7
DNA 96–8
double blind study 22, 176
Drake, Stillman 60–1
dummy variables 260

effects of the factors 181
Einstein, Albert 77
electricity 49, 98–101
electromagnetic theory 24, 72, 75, 76
electromagnetic spectrum 49
electromagnetism 72, 75
electrons 98–101
engineering 66
epidemiology 87
error 128, 129, 237–8, 243
error propagation analysis 247
evolution 22, 65, 70, 78, 85, 105, 107
expectation 176
experimental design 175, 230
    continuous factors 195–207
    design resolution 190–1
    determining confounding pattern 188–90
    factor screening 187–8
    importance of interactions 186
    one categorical factor designs 176–8
    screening designs 191–5
    several categorical factor designs 178–85
    statistics and the scientific method 175–6
experimental method 3
experimental treatment 140
experiments 62–3, 83
    assessment 108–9
    nature of 83–4
    questions and answers 85–8
explanatory theories 74–5
external noise 248–9

F-distribution 164
factor screening 187–8
factors 159, 176–207, 230, 236, 237, 250, 258
falsificationism 17
Faraday, Michael 34
FDIST worksheet function 158
Feynman, Richard 71, 78, 90–2, 101
fractional factorial designs 187–8
free fall 56, 58

frequency distribution 140
full factorial design 180, 192

Galen 88–9
Galilei, Galileo 21, 45, 46, 53, 54–62, 95, 219, 227
Gauss, Carl Friedrich 124
Gaussian distribution 124, 127–8
general knowledge 67
generalizations 69
genes 96–8
graphical analysis of data 256–9
gravity 45–6, 47, 57, 67, 76, 77
Great Scientific Revolution 41
Greek civilization 34–8
Greenfield, Tony 196
guesstimates 227

Hadamard matrices 195
half factorial design 187
Halley, Edmond 46–7
Harvey, William 88–90
heart 88–90
helium atoms 35
heritability of characteristics 93–6, 105–7
Hershey, Alfred 96–8
histogram 122, 127
Hubble, Edwin 85
Huygens, Christiaan 58
hydrogen atoms 35–6
hypothesis 17–19, 22, 23, 71, 86–8, 113, 141–2, 146
hypothesis-driven study 87
hypothesis-generating tools 222–6
hypothesis testing 23, 139–64
hypothetico-deductive method 16–19, 21–2, 72

inclined plane 55–66
independent samples 150
independent variables 121
inductive method 14–16, 23–9, 88, 141
input–output diagram 242–3, 248
interaction effect 182, 183–4
internal noise 249
Ishikawa diagram 222, 241, 242

Kelvin, Lord 235
Kepler, Johannes 16, 19, 21, 25, 42–5, 49, 71, 74, 77
Koyré, Alexandre 56–8
Kuhn, Thomas 78

Lavoisier, Antoine 27–8
Laws of Nature 66, 71
Leaning Tower of Pisa 54
lift-off length 72–5, 216–18, 242, 244, 249, 255, 256, 257–8, 259

linear combination 178
LINEST worksheet function 197–9
logic 14

magnetism 49
main effects 182
Maslow, Abraham 77
mathematical analysis of data 259–60
mathematical law 55, 56
Maxwell, James Clerk 49, 72, 75
mean 117
measurement system analysis 159–63, 244–8
measurement uncertainty 236–8
measures of central tendency 117
mechanics, Newtonian 19, 20, 27, 47, 74
mechanics, quantum 20, 29, 48, 74, 219
median 117
Mendel, Gregor 27, 93–6
Mersenne, Marin 56, 57–8
Middle Ages 38–47
Miletus 35–7
Millikan, Robert 67, 99–101, 108–9, 262
mindset 6
MMULT worksheet function 190
modal bin 122
multi-objective optimization 205
multiple linear regression model 170
multivariate data analysis 259–60
Musculinus, Mark 102
music and mathematics 39
Mysterium Cosmographicum 43

naive inductivism 14
nature of science 11
    confirmation, role of 21–3
    consequences of falsificaction 19–21
    hypothetico-deductive method 16–19, 21–2
    inductive method 14–16
    perception is personal 23–9
    scientific approach 11–14
    scientific community 29–30
Newton, Isaac 16–17, 28, 45–7, 76
Newton's laws of motion 71
Nobel Prize 21, 67, 68, 78–9, 98, 100
noise 129, 237, 248–9
normal distribution 122, 124, 129
normal plot 147
normal probability plots 129–32
normal probability scale 130
normal score 131
NORMDIST worksheet function 125
NORMSDIST worksheet function 126
NORMSINV worksheet function 131
null hypothesis 141, 142, 146

observation 15, 24–6
observation statements 16

observational studies 7, 40, 71, 85–6, 87, 108, 181
Occam's razor 74
On the Revolutions of the Heavenly Spheres 41
one categorical factor experimental designs 176–8
one-sided tests 142–3
optics 16–17, 45, 74, 75
optimization 205
orthogonality 178–9
    check 189
outliers 125–6
oxygen 27–8

p-values 164–5, 202
palcebo 87
papers, writing 260–4
parsimony 74, 167, 171
passive observation 7, 84–5
Pearson's correlation coefficient 165
peer review process 9, 13, 26, 30, 70, 264
pendulum 57–8
perception 23–9
periodic table of the elements 71
Ph.D. studies, undertaking 5, 65–6
    novel research 218–21
Ph.D. thesis, writing 264–5
phenomenological theories 74–5
philosophy of science 13–14, 30
phlogiston 27–8
photosynthesis 4
physics 19, 29, 39, 40, 41, 45, 47, 54, 65
Pisa 54
placebo 177
Plackett–Burman designs 187, 195
Planck, Max 219
planning research 211
    checklists 229–31
    hypothesis-generating tools 222–6
    scope 221–2
    thought experiments 227–9
    three phases 211–13
plant breeding 93–6
plate tectonics 70, 85
Plato 38, 40
PLOTMATRIX worksheet function 257
polarization 26–7, 214
pooled estimate of the variance 151
Popper, Karl 17
population standard deviation 119
populations 115
power analysis 148–50, 246
precision 116, 128, 244
precision improvement 246
prediction 20, 22, 74, 168, 171, 203, 206
Priestly, Joseph 27–8
Principia 46

probability density functions 123–4
probability distribution 122–6
probability points for *t*-distribution 270
process diagram 225
Ptolemy 19–20
Pythagoras 38–9
Pythagorean cosmology 39, 40
Pythagorean solids 39, 43, 44, 49, 97
Pythagorean theorem 39, 120–1

quantum mechanics 20, 29
quarter factorial design 187
QUARTILE worksheet function 132
quartiles 117

R^2 value 170, 202
RAND worksheet function 185
RANDBETWEEN worksheet function 185
random error 237
random sampling 122–3
randomization 141–2, 243
range 117, 230
Rayleigh scattering 4, 213
relative accuracy 246
religion 35, 36–7, 38, 44, 45
research areas 219
research characteristics of 68–70
research questions 86, 219
research studies 4–5
residuals 118, 119, 161, 167, 168, 169–71, 206,
    207, 257
resolution of a design 190–1
response surface 196
response surface designs 195–207
response variables 230
rotable design 200
Russell, Bertrand 15, 37
Rutherford, Ernest 99

Samos 38
sample standard deviation 119
samples 115
scalars 178
scatter plot matrix 257
scientific approach 11–14
scientific community 29–30
scientific method 1, 3, 6, 30, 175–6, 212
scientific papers, writing 260–4
scope 221–2
screening 187
screening designs 191–5
Semmelweis, Ignaz 18–19, 28
Settle, Thomas 58–60, 63
several categorical factor designs 178–85
sexual selection 105–7
significance 141, 146, 164

significance level 141
skewed distribution 122, 130
skills 6
spacetime 77
specific knowledge 67
stability check 249
standard deviation 118, 125, 126, 128, 132,
    134–5, 150, 162, 246, 247
standard error of the mean 247
standard normal distribution 125, 126
standard normal probabilities 269–70
statistically significant result 141, 146, 164
statistics 113
    central limit effect 126–8
    confidence intervals 132–4
    data analysis 113–15
    descriptive statistics 116–22
    normal probability plots 129–32
    populations and samples 115
    probability distribution 122–6
    t-distribution 134–6
statistics for experiments 139
    analysis of variance (ANOVA) 155–9
    comparing two samples 150–4
    correlation 165–6
    interpreting *p*-values 164–5
    measurement system analysis 159–63
    one-sided and two-sided tests 142–3
    other hypothesis tests 163–4
    power of a test 148–50
    randomization 141–2
    regression modeling 167–71
    t-test for one sample 143–8
STDEV worksheet function 132
Stokes' law 99
straight line equation 167
sum of squares 119
systematic error 237–8, 243

t-distribution 134–6
    probability points 270
t-test
    one-sample 143–8
    paired 177
    two-sample 150, 153, 154, 158, 176
TDIST worksheet function 152, 199
testability 22–3, 67, 69, 70, 74
Thales of Miletus 35–6
theoretical concepts 71
theory and reality 75–7
theory building 70
theory-dependent observation 28
thermodynamics 48, 66
thesis writing 264–5
Thomson, J.J. 98–101
thought experiment 46, 47, 54, 89, 224, 227–9

three phases of research 7, 8, 211–13
three-factor half factorial design 192
TINV worksheet function 143
trends in data 240
two-level factorial design 188
two-sided tests 142–3
type I error 148, 149–50
type II error 148, 149–50, 176

unbiased estimate 120, 157
unburned hydrocarbon (UHC) emissions 102–5
uncertainty 114, 236–8
uniform acceleration 54
unpolarized light 214

variance 119, 151
variation 116
vectors 178

Waring blender experiment 96–8
Wegener, Alfred 85
why-analysis 220
widowbird sexual selection 106–7
writing Ph.D. thesis 264–5
writing scientific papers 260–4

Yates order 180

Zola, Émile 63